FOLLOWING FIFI

FOLLOWING FIFI

My Adventures Among Wild Chimpanzees:
Lessons from Our Closest Relatives

JOHN CROCKER, MD

FOREWORD BY JANE GOODALL

PEGASUS BOOKS
NEW YORK LONDON

FOLLOWING FIFI

Pegasus Books Ltd.
148 W. 37th Street, 13th Floor
New York, NY 10018

MAIN PLAYERS IMAGES
Page xviii, Fifi and Freud. Photo by John Crocker, 1973.
Page xx, Figan. Photo by Grant Heidrich, 1974.
Page xxi, Passion. Photo copyright © The Jane Goodall Institute by Derek Bryceson.
Page xxii, Prof. Photo by John Crocker, 1973.
Page xxiv, Melissa and Gremlin. Photo by John Crocker, 1973.

LINE DRAWINGS
Page xxvi, Map of study site location in Africa. Illustration by Shayna McDonnell,
graphic designer, United Creations, Seattle.
Page xxvii, Author's hut in the Gombe forest, 1973. Illustration by Phil Keane,
artist and illustrator, Seattle.

First Pegasus Books edition December 2017

Interior design by Maria Fernandez

Library of Congress Cataloging-in-Publication Data is available.

ISBN: 978-1-68177-568-5

10 9 8 7 6 5 4 3 2 1

Printed in the United States of America
Distributed by W. W. Norton & Company
www.pegasusbooks.us

This book is dedicated to Jane Goodall
for her heroic daily accomplishments—
delivering a message of hope and a sense of purpose to young and old alike.
Over her lifetime, she has inspired millions of people across the globe
to take part in efforts to make our environment a better place
for all living things.

Jane will be happy to know that this book is also dedicated to Fifi,
another mother who furthered our understanding of primate parenting
and the value of patience and reassurance in raising our young.

CONTENTS

FOREWORD

John Crocker has written a truly fascinating book. As I read it I was transported back to those wonderful days when I lived in Gombe with my son, "Grub," when he was a small child, in those far-off days when bright, highly motivated, and carefully selected undergraduates from the Human Biology program at Stanford University came to help with the collection of data. Many of them, like John, were premed.

It was fascinating to me to watch how the Gombe experience—the chimpanzees and baboons, the forest, the close contact with a different culture in the form of the young Tanzanian field staff—affected the new students when they first arrived. And I loved watching how they changed. Some were a little startled when they encountered the primitive nature of their accommodations, even though we had tried to prepare them. Some were nervous to start with; others seemed to take everything in their stride. John was quiet, a little shy and reserved. He was very thoughtful, and I could see him take everything in, processing the new information.

The research project assigned to him was to follow four mothers, Fifi, Melissa, Passion, and Nova through the forest and collect information on their interactions with their infants. As the weeks went by, I came to appreciate that John was a really good observer, not afraid, as some were, to use his intuition when trying to understand why the chimpanzees behaved as they did.

I enjoyed talking with him about his ideas. In fact I appreciated those discussions more than he realized. I had little time then for actually observing the chimpanzees, for I was fully occupied with trying to run the research station, write up data—and be a good mother! I sometimes felt a bit lonely, sort of cut off from the camaraderie of the student group. John was one of the few with whom I felt I had a meaningful relationship outside that of teacher and student.

That John would make an excellent family doctor was obvious from the start, for he was always helpful, sensitive, and caring, but it was only much later—especially after reading an early draft of this book—that I realized just how profoundly his Gombe experience has influenced him. His patients have surely benefited from his understanding of the human condition, gleaned during his long hours with our closest relatives in a remote forest.

Being a good doctor does not necessarily mean that a person will be a good writer, but John is both. He describes vividly his experiences in the forest, his growing closeness to the chimpanzees, and all that he was learning about them. He shares too his evolving friendship with one of the young Tanzanian field assistants with whom he spent so many hours in the forest, from whom he learned so much, and who, as John discovered years later, learned so much from him. He weaves all these experiences seamlessly into an account of his subsequent experiences as a busy medical practitioner.

It is fascinating to read how John's observations of chimpanzee children with their mothers and other family members helped him understand the problems of a human child. How understanding the motivation behind the dramatic dominance displays of young male chimpanzees helped him pinpoint the possible cause of unruly behavior in a young human boy. Both the chimpanzees he studied and the Tanzanians he came to know helped John appreciate the importance of community, the emotional support of friends, and the need of our infants to form close and affectionate bonds with one or more trusted adults.

And finally there is the importance of family and community in the lives of chimpanzees and humans, and John's relationship with his own family. An important part of the book is the trip he took with his teenage son, Tommy, back to Gombe, hoping it would be a good experience for both of them. Would Tommy understand why this tiny national park with its human and animal inhabitants had been so important to John? The description of the father sharing with his son his love of the place where he had learned so much is moving, as is the reunion with Hamisi, the Tanzanian with whom he had formed a friendship that withstood the test of time.

John, I want to thank you for writing this. I have loved reading your accounts of your hours with Fifi and Freud, Melissa and Gremlin, Figan and Mike. It has taken me back to another era—before I started my life on the road as a conservationist, before you were plunged into the sometimes-unbearable workload of the family doctor in today's world. Not only did it take me back to Gombe, but also reading it made me reflect on how Gombe influenced my life too. I am glad we shared some of that time together, and I am glad that our friendship, too, has lasted over the years.

—Jane Goodall

PREFACE

In June 1973, as my Stanford classmates readied for graduation, I embarked on an eight-month sojourn to study Dr. Jane Goodall's famous chimpanzees—research that would forever change my view of the world and of myself. My time with Dr. Goodall and the Gombe chimps would also influence my approach to both fatherhood and the practice of family medicine in the years to come.

Beyond the wisdom I gained from studying the chimpanzees, I also enjoyed experiencing the ten-mile stretch of Tanzania's Gombe Stream National Park along the shores of Lake Tanganyika, which was nature at its most spectacular. My memories of flower and fruit fragrances at the beginning of the rainy season and magnificent sunsets over the lake and beyond the rolling Congolese hills remain vivid. I can still recall the night sounds of bush pigs rustling and cicadas buzzing as I fell asleep in my thatched hut. Gorgeous birds and astonishing insects would suddenly appear, inciting curiosity and wonder. Stealthy green-and-black mamba snakes would occasionally catch my eye, and I delighted in watching ancient fish species darting through the clear waters of the lake. On cloudless nights, the brilliant stars astounded most newcomers, who weren't used to seeing the Milky Way look so intimate with the earth.

I captured impressions of the wild forest and my days with the chimps in the scores of letters I wrote to my parents from Tanzania and in my journals and field notes. After I returned home, completed my medical training, and began working long hours as a family physician while also raising a family in Seattle, I often reflected on my time at Gombe.

Almost daily, I still think about the chimpanzees and Jane Goodall's contributions to our understanding of primate behavior. I have incorporated lessons I learned from the chimpanzee mothers into raising my own sons and giving advice to my patients. I've internalized the importance of a strong emotional attachment, physical contact, and reassurance, which young chimps need as crucial elements in their development. These lessons remind me to be as patient and engaged in my relationships with my sons and my patients as the chimp mothers were with their offspring. In particular, I reflect on how Fifi, the chimpanzee matriarch, related to her son Freud.

When I witness the curious exploration of my young patients rocketing around the exam room, I remember Fifi's acceptance and tacit approval of Freud's uninhibited and joyful behavior. Her example helps me incorporate tolerance into my child-rearing and medical practice. At the same time, I remember that whenever a dangerous situation arose in the African forest, Fifi would forego patience and scoop up Freud, clutch him tightly, and make a quick escape. Our instincts as parents are reflected in this kind of deeply human primate behavior. When at age two my eldest son wandered over to the edge of a large fishpond while I was trying to take his picture, I instantly dropped the camera, shouted, and grabbed him. My response was the automatic reaction of a parent, and I felt at one with Fifi at that moment.

In my medical practice, I find it helpful to view common conditions such as anxiety and ADHD in my patients from an evolutionary perspective. Knowing the close similarities between chimp and human DNA, I can reflect back on Frodo, an alpha male chimp with ADHD characteristics, and understand why these traits allowed him to successfully serve his community and why similar traits may persist in humans today. Over time I've realized even more how an individual's unique genetic makeup is important in understanding their depression and anxiety.

The influence that the Gombe chimps had on my personal and professional life was so strong that after thirty-two years I felt compelled to pull out those letters my parents had carefully saved and use them as a foundation for sharing my story.

Part one of this book is a chronicle of my first Gombe trip at age twenty-two. Part two presents clinical cases from my practice that demonstrate how my time with Jane and the chimps influenced my work with patients with depression, anxiety, and other medical conditions. Finally, part three is an account of my emotional return to Gombe thirty-six years later with my twenty-year-old son, Tommy. It recounts my reunion with Jane, the chimps, and some of my former field assistants. Along the way, I share how my experiences with chimps in the wild have influenced my approach to raising my children and helped shape my worldview. Throughout the book, I highlight my relationships with Jane, my son, my Tanzanian field assistant, and the chimps.

My experiences at Gombe laid an unexpected foundation for my life as a physician and father and inspired me to add another voice in support of protecting chimpanzees and their environment. Spending time with these extraordinary animals in their own remarkable world has helped me understand more about what it is to be human.

THE
MAIN PLAYERS

FIFI

Fifi followed in the footsteps of her famous mother, Flo, the charismatic matriarch whom Jane had studied closely in her early years at Gombe. Flo had given birth to Fifi just before Jane's arrival at the camp. When I arrived at Gombe, Fifi was age fifteen, and she was noticeably both confident and playful as a mother. I often saw her tickling and rolling with her son Freud, who would laugh and come back for more. I loved Fifi's nonchalant yet effective manner in raising not just Freud but also her eight other offspring! She was strong and energetic. Unusual for a lone female, she was observed single-handedly hunting and killing a small bushbuck for food. Even the researchers found Fifi, who was well respected and well integrated into the Kasekela community of chimpanzees, engaging. Of all the chimps in Kasekela, Fifi taught me the most about the importance of patience, reassurance, and confidence in parenting.

FREUD

Freud played with everyone and everything. At two and a half he was more competent and confident than most males his age. With his superior gymnastic capabilities, he swung and leaped across branches high in the trees. He would do pirouettes when scampering after his mother, swing from low-hanging branches, and then race to catch up. He was known to arouse an entire baboon troop and then escape up into a tree while four or five large baboon males waited at the base. From an early age, he displayed singular dexterity, confidence, and fearlessness. Freud was destined to become an alpha male at Gombe.

FIGAN

Figan was Fifi's older brother and the alpha male of the community. Very suave, yet strong and competent, he displayed just the right amount of aggression and drama to maintain his high position. Highly respected, lean, and smart, he seemed less showy than some of the other high-ranking males. His older brother, Faben, often supported him and helped him to attain and maintain his high rank.

PASSION

Not the most attentive mother, perhaps because of her strong drive to seek food, Passion was less responsive to two-year-old Prof than other chimp mothers were to their offspring. When Prof whimpered, Passion would not always respond, especially if she was busy eating fruit high in a tree—even when he was close by. A bit of a loner, she did not socialize much with the other chimps and displayed dominance around other adult females. Passion and Pom (her oldest offspring) did the unthinkable—they cannibalized at least three infants in the community over a two-year period.

PROF

Prof loved to do acrobatics while riding on Passion's back, but when Passion was nervous or distracted or simply inattentive, he was markedly less adventuresome. Perhaps because of this, Prof was very attentive to his mother when he sensed she was getting ready to move to a different location. He kept close to her, especially on her more distracted days. Luckily, Prof had a big sister, Pom, to play with, and she made up for his mother's lack of playfulness.

MELISSA

A quiet, thoughtful, and cautious parent, Melissa was mother to ten-year-old Goblin and three-year-old Gremlin. Melissa would hang out in groups but not engage in as much grooming and other activities as some of the other mothers. She was very patient with Gremlin, once waiting thirty minutes for her to finish a play session before moving on with her. She and Gremlin were a well-coordinated pair with a complex set of subtle communications between them.

GREMLIN

Gremlin was quick to learn the crucial skills of termiting and nest building. She quietly observed her mother, Melissa, for long periods and with great focus. One of her unique characteristics was her interest in carrying around objects such as Strychnos fruits, flowers, or even an old shirt she found near camp. She was also observed drumming on a hollow buttress of a tree, an activity usually reserved for adult males. Another unique behavior was the way in which she would initiate her own retrieval by Melissa. Gremlin would extend her hand to Melissa in times of great fear, signaling that she wanted her mother to pick her up. This was highly distinctive. Other chimps would usually just whimper and then approach their mothers during such times. Gremlin was confident and loved playing with Freud. She later became a very successful mother, and she was the first chimp in the wild documented to raise twins to adulthood.

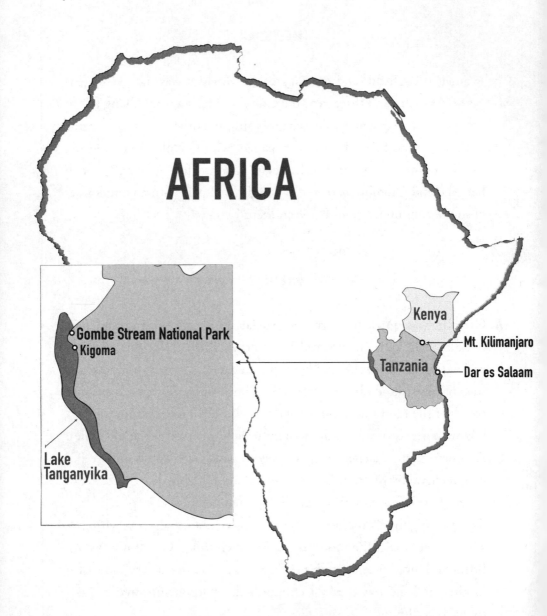

Map of study site location:
Gombe Stream National Park

My hut

FOLLOWING FIFI

PART ONE

INTO THE FOREST

CHAPTER ONE

GOMBE ARRIVAL

June 1973, Gombe Stream National Park, Tanzania, East Africa

I sat spellbound, leaning against the sturdy trunk of a tall, graceful palm-nut tree. The beauty and foreignness of the Tanzanian forest around me had me transfixed. I felt alert and yet strangely calm in this primal setting.

With me was my field assistant, Juma, who spoke only Swahili, and whose very rich black skin vividly contrasted with my light complexion. Juma was steady, confident, and thoroughly knew the chimps and the land. Though I'd known him only a few weeks, we had connected and had become good friends as he expertly led me through rugged valleys to follow and observe a community of fifty chimpanzees.

There were no evergreens or frozen ground around us like there were back home in Minnesota. The air felt warm and humid. Amid the symphony of forest sounds, my mind was freed from practical concerns—no thoughts of filling the car with gas, going to the bank, buying groceries, or meeting a friend for lunch. In fact, I was barely conscious of thinking at all. I had quiet time to gaze at a brilliant red forest flower and daydream about the past. In broken Swahili, I tried to share a few personal stories with Juma as we waited in the afternoon shade for twenty-one-year-old Figan, head male of the Kasekela chimp community, to come down from a nearby tree.

"Look!" Juma pointed to the tree where Figan effortlessly descended before heading in our direction. I was surprised by the blackness of Figan's hair and his sleek 85-pound muscular form. On all fours, with his shoulders higher than his powerfully bent hind legs, Figan walked toward us with complete confidence. He came so close we could hear his breathing, and then he passed right by as if we didn't exist.

Juma and I sat frozen, not wanting to draw attention to ourselves. I felt my heart rate speed up as he passed us; Figan could rip a man apart in an instant if he wanted to.

Figan stopped and turned his head toward a resonating pant-hoot from his brother Faben, which emanated from across the lush valley. Heartily, Figan returned the call.

Did a similar scene play out four million years ago, when early humans were emerging from an environment similar to this? Looking at Figan, I wondered what our early ancestors were like, how they survived in the forest, and how they related to their closest living relative, the chimpanzee.

Touching my arm and gesturing for me to come, Juma led me through thick vines and low-hanging branches for a few miles into another valley, where mother chimps and their infants had joined adult males to feed on newly ripened *mabungo* fruit, also known as milk apples.

These nutritious fruits, with fleshy sweet pulp that turned slightly acidic-tasting near the seeds, abundantly decorated the trees. I marveled at how the chimps could sit so relaxed high up on flimsy branches, which they grasped with hands and feet as they picked and ate the fruit. The smell of those two- to three-inch milk apples made my mouth water and my stomach rumble; my most recent meal was the peanut butter sandwich I had devoured six hours earlier.

In the early evening, as the sun began to set, the adult chimps carefully selected trees in which to build their nightly nests. Infants snuggled in with their mothers, and males gave low-pitched hooting sounds from their newly constructed beds to signal their location. A reddish-purple sky faded into twilight, and the forest was again quiet when Juma and I headed back to camp to join our human companions.

The opportunity that had led me to this remote part of Africa arose in the fall of 1971, during my junior year at Stanford. I enrolled in a primate behavior class in human biology, with Jane Goodall lecturing as a new visiting faculty member. I can still picture sitting in the second row of the three-hundred-seat auditorium, seeing the famed Dr. Goodall in person for the first time as she showed colorful *National Geographic* footage on a huge screen. I was mesmerized by the close-up shots of her sitting a few yards from the wild chimpanzee families she had studied for more than ten years. This graceful young Englishwoman was the first human to get so close to the chimp community. Seeing her in the film gently observing the chimps in their natural setting made me feel like I was right there with her.

Afterward, I could not stop thinking about the image of an older chimp, David Greybeard, signaling his acceptance of Jane during her initial time with the chimps. He did so by cautiously touching her hand as she reached out to touch his—an image that millions of people world-wide know, illustrating the bond between species. It's a tender reminder that we share 96 percent of our DNA with chimpanzees.

Images of Jane and the chimps remained fixed in my mind during the next two months as I tried to concentrate on coursework. Memorizing the Krebs cycle and learning organic chemistry drifted lower on my priority list while Jane and the chimps moved up. Even in the quiet cubicles of the undergraduate library where I could usually focus on tedious assignments, I felt restless when I thought about Jane's life observing Flo, the charismatic matriarch chimpanzee, raising her offspring, Fifi and Figan, in the wilds of Gombe.

Interested in child development, I was contemplating a career as a pediatrician or family doctor. Perhaps my motivation to enter this field involved a subconscious desire to better understand my own upbringing. I knew the tensions in my home might have affected my self-confidence and security during my teen and college years. Thus I was intrigued that a major focus of Dr. Goodall's research involved mother-infant relationships among the chimpanzee families in Gombe, and I applied for Stanford's student research program there. Few students had participated at this point, and only two would go every six months to work on the chimp study.

In an interview with faculty members as part of the application, I eventually relaxed as I expressed my fascination with how young chimps grow up successfully in the wild. To make sure they understood my future plans, I admitted, "I don't intend to pursue a career in anthropology. I plan to attend medical school after graduation." However, I chose not to reveal that, though I loved the outdoors, I had never even been camping before.

One day, while I was working on a chemistry assignment, a dorm-mate startled me by knocking loudly on the door and telling me I had a call on the hallway phone.

I went down the hall and answered.

"Is this John Crocker?" a woman's voice asked at the other end. With my mind halfway still on chemistry, I said it was, but what she said next captured my complete attention. "I'm calling to let you know that you have been chosen to participate in Stanford's research program at Gombe!" I leapt off the floor with joy. Before I had even hung up the phone, I began to picture myself with Jane and the chimps in the spectacular landscape I'd seen in her films. I knew I would witness the intimate bonding of mother chimps and their offspring right before my eyes. What I didn't know was that this primate training ground would give me remarkable insights that would serve me well in my later work as a family physician.

I had already spent a year volunteering at the Peninsula Children's Center in Palo Alto, working with autistic and schizophrenic adolescents. There I experienced a full spectrum of normal to very abnormal behavior in children. I had also read about rhesus monkeys that had been isolated even for a short time early in their lives and how profoundly this affected their social skills. Although the cause of the abnormal behavior in the monkeys was more environmental, compared to the genetically inherited disorder of the children at the center, both groups had similar behaviors that resulted in difficulty socializing within their communities. As I studied more about primate behavior, the similarities I saw between nonhuman-primate and human behavior were striking.

It was already clear to me from Dr. Goodall's films, including her first, *Miss Goodall and the Wild Chimpanzees*, that the chimps at Gombe acted a lot like humans. I was also beginning to realize that they might share a similar range of emotions. While watching footage of a young chimp responding to the death of his mother, I was struck that this eight-year-old, Flint, seemed to experience grief and loss like any human child would.

As his mother, Flo, lay lifeless next to a stream, Flint continually approached and touched her, then shrieked and backed away. He spent the next three weeks very close to her body, seeming quite upset. He didn't eat, and soon he died very near to where Flo lay. When this same chimp, at age five, was being weaned by his mother, he would throw temper tantrums similar to those of human children in their terrible twos. These types of behaviors drew my focus and spurred my thinking about how I had underestimated the capacity of chimps to feel emotional pain. I would learn much more from them before my Gombe education was through—thanks to Jane Goodall's extraordinary guidance and teaching.

Dr. Goodall's story is so well-known that it scarcely needs repeating. As a young child, she dreamed of going to Africa to study animals. She worked diligently after high school to earn enough money for a boat ticket to Kenya, where she met the renowned archaeologist Dr. Louis Leakey, who chose Jane to conduct the first-ever study of chimpanzees living in the remote Gombe forest along the shores of Lake Tanganyika.

Jane started her work with the chimps in 1960 with only the assistance of her mother, Vanne, and a Tanzanian guide. Jane's life grew so intertwined with her work in Gombe that her name became virtually synonymous with these chimps. Her revelatory research into their lives brought them to life for scores of people who would never see a chimp except in a zoo. Her work made people aware that chimps are close cousins of our human family—with personalities, affections, and behaviors that we could recognize as aspects of our own species.

With a year to prepare for the Gombe adventure, I spent my free time learning Swahili by listening to tapes at Stanford's language lab, and I ventured to Southern California to Lion Country Safari, where I studied chimps living on an island. A retired primatologist taught me the meaning of various chimp calls as we sat in a rowboat all day, watching the primates from a distance. I didn't mention my lessons in imitating chimp calls to my parents, who were having a hard enough time accepting my postponement of medical school. And before long, an unexpected situation arose that provided me with a far better learning experience than any I could have planned.

Enter Babu

Babu was a spirited orphaned chimpanzee. He had been lovingly hand-raised by a human couple, Joe and Janet Hoare. Babu was born in West Africa, only to be snatched away from his mother by poachers and put on sale as food at a local market. The animal-loving older American couple was on a three-month work assignment in Liberia and they agreed to take Babu to relieve another couple who had purchased him at the market in order to save his life. Joe and Janet repurposed a nice basket for Babu to sleep in, fed him formula for the remainder of their assignment, and then flew with him back to the States. Their rustic house in Woodside, California, became Babu's home. The couple devoted much of their time and energy to raising him, but by two years of age, Babu was growing more aggressive—as is natural—and he needed constant supervision.

Having heard from friends at Stanford that I would be studying chimpanzees with Jane Goodall, Joe and Janet assumed that I knew a lot about these primates and contacted me to see if I would be his companion.

Janet called me without anyone warning me ahead of time and explained, "We're looking for someone who could take our baby chimp one afternoon each week while we do our shopping and run errands. Since you are familiar with their behavior, you are our top choice for the position."

In reality, I didn't have a clue as to how to handle a young chimp. I was just beginning to learn in Jane's course about how young chimps are raised by their mothers in the wild, but I was still very ignorant about the dangers and practical problems of raising a captive chimpanzee in a human community. A chimp's physical strength alone could cause great harm to a caretaker, especially as the animal matures. Regardless, the couple proposed that I take Babu on field trips on afternoons when I didn't have classes, and I agreed.

I was introduced to Babu in his home. When I entered and greeted Janet, Babu was vigorously and repeatedly leaping from the sofa to the table to the floor and then to the kitchen counter. As I watched in amazement, I questioned whether or not I could handle this hairy ball of energy on my own. Yet as soon as we left the house together for our first outing, his focus turned almost completely to me.

Every week Babu and I enjoyed a new adventure, visiting parks and nature trails. I realized that Babu was capable of aggression toward strangers, especially a small child who might threaten him by cornering him, but I soon learned that he was frightened of new people or animals he encountered and tended to avoid them. He kept me in his sight for reassurance. Though I was also initially worried that I might lose Babu if he were to run off, I discovered that he was so attached to and dependent on his caretaker that he kept track of me at all times.

Babu traveled in the front seat of my dark green '63 Mustang and peered out the window at passing cars and scenery. Stoplights were always distracting, as people in cars next to mine were clearly startled to see Babu staring at them from the passenger seat.

Like young chimps in the wild, Babu still needed milk in his diet, so I carried a few bottles of cow's milk with me, along with apples, bananas, and figs. He was cooperative enough to let me change his diapers when needed—and luckily for me they were the disposable type.

Whenever we'd visit one of Babu's favorite city parks, parents would stare at my hairy companion wearing Pampers. Babu loved to leap across the playground equipment, then dangle from one arm as he surveyed his surroundings. Because he preferred to be far above the ground, his area of play didn't disturb the children twelve feet below.

If frightened by a dog or shrieking child, Babu would run to me, lunge into my arms, and hug my torso with all his might. He would then continue to cling with his power hug while I strolled around the park until he was reassured enough to go back to his magnificent aerial acts. When he was in my arms, children sometimes approached, curious about the hairy primate I held, but most youngsters in the park were having too much fun with one another to pay much attention to him. One day, a father came up to me and said with a smile, "Hey, that sure is an ugly kid you got there!" I smiled back with great pride.

People were curious about our relationship, wondering why a twenty-year-old college student would be in a park with a young chimpanzee in diapers. Even I wondered at times how people would view us, but this was the early 1970s and California, after all. However, despite how strange our relationship might have seemed to others, it was with Babu that I first began to experience strong feelings of fatherhood. I enjoyed filling that nurturing role as well as just having fun watching Babu climb trees and explore his surroundings.

One day, as it grew close to the time for my departure to Africa, Janet approached me. Babu and I had just returned from a walk, and I was already feeling a bit sad at the thought of leaving him.

"John," Janet said softly, and I immediately felt worried. "John, I need to tell you something. When you get back from Africa, Babu isn't going to be here in our home." I was not surprised, although I realized I had been comforting myself with thoughts of coming back to resume our unusual friendship. She continued. "Babu is getting too aggressive, and we just can't keep him here anymore. He needs other chimpanzees and a more suitable home."

Babu would be moved into a large outdoor chimp compound near the Stanford campus, where he would be integrated with other chimpanzees. To do so successfully, he would have to give up his human contacts. I knew they were doing the right thing; it was the only reasonable alternative for him as he naturally became stronger and more difficult to contain in a home.

When the time came to journey to Gombe, saying good-bye to Babu was the hardest part of leaving. I knew that I would never again be able to interact so intimately with him, and that made it a thousand times harder.

We were at his Woodside home, and his caretakers had just returned from their errands. Babu was still showing off for me by leaping up and swinging on the cupboard doors. I grabbed my jacket, and he immediately lunged into my arms. When he hugged me more tightly and longer than usual, I wondered if he somehow sensed my sadness. I fought back the tears that burned my eyes, knowing I might not see him again.

I was concerned about how Babu would adjust to living with the other chimps. Also, I felt like I was abandoning him. The irony wasn't lost on me that he would soon be placed in a fenced compound at Stanford while I ventured into an African forest to learn what his life would have been like had he not been taken away from his mother in the wild.

I treasured my friendship with Babu but knew that my role in Tanzania would be utterly different. I would need to strictly observe scientific protocols and never interact with the forest chimps. I would have to give up the power hugs.

My emotional departure from his home that day gradually set in motion a feeling of quiet contemplation and a diminishing enthusiasm for some of the things around me. It was as if I had been caught in a slack tide—when the ocean waters stop coming in but before they go back out. It was a time of being in between. I was departing from Babu, my colleagues, and the comfort of the campus, and yet I still had a few weeks before embarking on my next journey, to the other side of the globe, to an area I knew little about.

As I drove back to my dorm, I stopped at the hillside of golden grasses and scattered oak trees where I had enjoyed late-afternoon jogs and picnicking with friends, and where I had sought out a quiet spot in which to write a biology paper my freshman year. I loved the peacefulness and natural beauty of this spot. I felt sad to know I might not enjoy it again. I parked the car at my dorm and walked across campus as the afternoon sun accentuated the red-tile roofs. I recalled fun times talking to friends and just sitting by the "The Claw" fountain and pondering life.

For my first three years of college, I had always anticipated celebrating my graduation with my college friends and then going off to medical school. Instead I would be leaving this phase of my life prematurely, before graduation. For the next few weeks, I felt like an outsider, even as my excitement for my unexpected life detour was nervously building. With some loneliness and tears of sadness as I said goodbye to my close friends at a farewell party, I gathered my packed belongings, and headed to the airport.

The Journey Begins

My journey to Gombe began with a twenty-four-hour flight (including a stop in Amsterdam) to Dar es Salaam, Tanzania, where I met Lisa, another student I had met briefly in orientation meetings, who joined me for the dusty, bumpy eight-hundred-mile train trip across the Serengeti to Kigoma, on Lake Tanganyika.

We stayed glued to our windows for the duration of the trip, spotting herds of wild zebras and an occasional ostrich in the distance as we traversed endless deserts and savannas—and the sunsets were spectacular.

I wrote to my parents:

> Lisa and I got the last two tickets on the train, which was lucky, as the train doesn't run every day. The trip takes over forty-eight hours because the train stops at every little village on the way to Kigoma. At each stop, villagers would sell

machunguas (oranges) and even live roosters to the passengers by passing the items through the compartment windows.

With earthy, humid air blowing through the open train windows, aromas of spicy foods cooking, and colorfully dressed Tanzanians moving down the aisle, I knew I wasn't in California anymore. Joyful curiosity and pleasure in the captivating new culture percolated in my mind. At night I would awaken to the occasional sounds of infants crying in neighboring compartments or the smell of fires from the intentional burning of fields.

Having Lisa with me on the journey allowed me to share my excitement and my insecurities about being in a foreign land. She won my heart early on because of a hair incident. That day, she had nodded off in her seat when two young girls with tightly braided black hair started touching her long, wavy blond locks. Usually serene and focused, Lisa looked rather startled when she opened her eyes to find the two girls gently stroking her hair.

"*Safi sana*," beautiful, they told her.

"*Asante sana*," thank you, Lisa replied, recovering—but then she went on to tell them in Swahili that she thought *their* hair was more beautiful than her own. That was Lisa in a nutshell: sensitive, thoughtful, and always thinking about others. We both felt an easy connection with the Tanzanians we encountered, especially when we tried to speak Swahili. Lisa's subtle but penetrating smile during these moments added a deeper bond to our growing friendship.

Lisa and I climbed off the train at its final stop, Kigoma, a bustling lakeside market town. With only a small suitcase and backpack apiece, we boarded a six-person, gas-powered boat driven by a park ranger for the three-hour trip along the shoreline of the immense Lake Tanganyika.

Exhausted to the bone yet tingling with anticipation, I peered intently from the boat as we pulled up to the shore of Gombe Stream Research

Center, just as Jane's even smaller boat had thirteen years earlier when there were no signs of civilization and she set up camp with Vanne to begin her research. The wild, densely forested Kasekela Valley, where Jane conducted much of her study, lay before us. I stared at the barren Rift Mountains that lay farther up, above the valley, forming the eastern border of the Great Rift Valley in East Africa. Along the lakeshore, pristine beaches welcomed us with lapping waves on the bright white sand. To the west, forty miles across Lake Tanganyika, were the hills of the Congo. Gombe Stream National Park, founded in 1968, ran ten miles along the eastern shore of the lake and two miles up to the mountaintops.

Stepping out of the small boat, I was amazed by how clear the lake water was. "After you get back from tracking chimps through the rugged forest all day, you'll appreciate this lake to cool your body and relax before dinner," a staff member said to us in slowed-down Swahili. He was right. The lake would also add a few terrifying moments to my African adventure.

Jane greeted us with a warm smile when we arrived at the only structure in sight, the thatched communal building where students and researchers gathered to eat dinner and discuss the chimps they'd observed during the day. She wore the same style of khaki shorts and sandals that she wore in the *National Geographic* movies and looked just as serene and sincere. With warmth in her twinkling eyes, she said, "I'm glad you both made it safely." As I exchanged hugs with the renowned scientist, I felt like I was in a movie or dream, watching it all take place before my eyes. At the same time, I sensed that I would become part of the Gombe family and that we would all look out for one another in the coming months. I felt at home already.

Surveying the area near us where Jane had first set up her campsite in 1960, and then looking east to the forest edge, I imagined I could see David Greybeard, the first chimp who had trusted Jane enough to allow her to observe him closely. He had lived and died here, his life made famous by Jane's observations. Fifi and her chimp brothers Figan and Faben were still here, perhaps in the same valley. I was struck by the enormity of human history—from our evolutionary roots until now. Despite feeling like such a small speck in this grand scheme, I briefly shivered as I realized it all fit together beautifully! And along with the thrill of being at Gombe, I was humbled by the fact that I would be working alongside such a famous scientist.

That afternoon, Lisa, a field assistant named Esilom, and I hiked from the beach camp to the upper camp, twenty minutes higher into the valley. From the trail, we noticed tall grasses, tangled vines, thick bushes, and massive trees—some with leaves two feet wide. Esilom let us know we would be tramping through that type of terrain at times and that the plum tree thicket and the waterfall would be challenging and might require acrobatic talents to navigate. He joyfully enacted swinging across the stream at the base of the waterfall by grabbing hold of a four-inch vine nearby, and my stomach tensed a bit. Luckily, chimps, baboons, and human researchers had created natural pathways that allowed us some upright trekking.

I reflected back on my childhood, picturing my mother standing far below us, hollering at my friends and me, who were swaying in the tops of the trees in our backyard, ordering us to come down. Later, in high school, my two brothers, my sister, and I hooked up a rope trapeze that swung across a ravine. We played on it every day, building up our arms. I hoped those experiences might help me feel more confident in the Gombe forest. I guessed time would tell.

Soon after our arrival at the upper camp, several other Tanzanian field assistants, a few of the European researchers, and I gathered at a grassy area among the palms and talked about the state of the forest this time of year.

June was the end of the wet season, so the flowering plants and edible leaves were at their peak of growth. Wonderfully flavorful *mabungo* fruit was ripening throughout the valleys. Blue monkeys swung through the treetops while Cape buffalo and the occasional leopard roamed on the ground. Although confronting a leopard sounded terrifying, no one at the camp had seen one in years. Herons, crowned hornbills, and great spotted cuckoos might be seen in the ficus and tall oil-palm trees. The seventy-five to eighty degree temperature seemed about right for my attire: khaki shorts, a cotton shirt, and plastic sandals. This was so much more exhilarating than a classroom setting, and I could hardly believe that I would receive academic credit for my two quarters of "African Field Study."

A researcher named Bill told me, "You'll get hooked on this place." Bill was tall, with thick black hair and a full beard. He seemed completely relaxed and happy living at Gombe while doing his chimp research. He fit seamlessly into the scene of humans and animals living close to one another, partly because he was so hirsute.

Peering around at the lush jungle, I imagined Jane herself, thirteen years earlier, arriving here and conducting her study. She'd had little knowledge of what to expect. The *National Geographic* films had showed Jane in a similar setting, only higher in the forest, looking down from "the Peak" at the valley and searching with her binoculars for signs of chimp life. I was in awe of the fact that she had slept on the Peak alone, without even a tent, completely exposed to the elements. "These were some of my happiest moments," she had revealed during one of her lectures at Stanford.

Enjoying the moment, I tried not to think about the myriad of tasks ahead of me and the things I had to learn in the coming few weeks. For example, I had just met Hamisi and Yasini, two of the twelve field assistants when another field assistant also introduced himself as Hamisi. "*Hujambo*, Bwana *Johni*," he said—Hello, Mr. John. When I looked confused, he explained he was Hamisi Mkono and not Hamisi Matama, whom I had met earlier. In addition, I knew I would soon need to distinguish individual chimps, such as the two- and three-year-old Skosha, Prof, Freud, and Gremlin. I would record their behaviors and memorize their faces, body language, and sounds. It seemed daunting.

During this orientation, my self-confidence waxed and mostly waned as I began sizing up the disparity between the eloquent speech and experience of the four doctoral researchers and mine. They hailed from Cambridge and Oxford, and some were already published authors. In contrast, the casual and less mature nature of the three other Stanford students brought me solace; I identified with their playful and sometimes goofy behavior, though we were all serious about our work. And despite my initial perception that the English and Scottish accents of researchers around me projected more sophistication, this feeling rapidly melted away as they genuinely welcomed us with open arms.

And I had to adjust to Africa itself—the heat, the smells, the stripped-down outdoor living. I would live alone in one of the huts scattered a half mile from the beach camp—each solitary hut out of sight of the others. Sleeping alone in the dark in an African forest seemed an intimidating prospect.

To my amazement, on that very first day at Gombe, several chimps calmly walked into camp while the researchers and field assistants were briefing me about my role on the study team. The first chimp I saw was Figan, the top-ranking male and son of the former matriarch, Flo, who had died a year earlier. Figan arrived with his brother Faben and the huge Humphrey; they swaggered into camp with powerful stances and serious facial expressions.

The sight of these strong, beautiful animals stunned me. They seemed to study their surroundings, paying no attention to their human observers. When they came so close that I could catch their unfamiliar scent, I couldn't help but startle a bit. We were very quiet and moved slowly to avoid interfering with their behavior. We were united in our strong belief that this forest was theirs and that we were merely visitors whom the chimps allowed to coexist with them. Our only job was to watch and record. We all grew to appreciate the long months Jane had endured before the chimps finally allowed her to get close enough to study them. Now observers had become just part of the landscape to the chimps.

The thrill of meeting members of the Kasekela chimp community—the fifty chimps Jane came to know at Gombe—soon overshadowed my worries about living in the African forest. I felt lucky to know I would observe Flo's offspring, whom we called the *F* family, consisting of Figan and Faben and their sister Fifi. Fifi was fifteen years old and the mother of two-and-a-half-year-old Freud. Two families I studied included older siblings. Adolescent Goblin still traveled with his mother Melissa and sister Gremlin; likewise, Pom with Passion and Prof. Some of the roaming young adult males, including Satan, Jomeo, and Evered, also added character to the fascinating and engaging community of chimps.

A few miles to the south lived the smaller Kahama community of chimps, also named for the for the valley they inhabit. They interacted on occasion with the Kasekela community, but another group to the north stayed completely separate. This made for a total of about 160 chimpanzees within the steep valleys of Gombe Stream National Park. The vegetation of this twenty-square-mile park ranges from grassland to alpine bamboo to tropical rain forest. This was the chimps' world. Now that I was here, Jane's original campsite and the chimps she wrote about became alive and real for me. Africa was alive and real for me. I would begin to follow these extraordinary creatures—and begin to learn from them—early the next morning, my second day in camp.

Later that evening, I ate what I could of the fish, rice, and beans prepared by our Tanzanian cook. My gut was battling an introductory bug, manifested by a slight fever and lack of appetite, but I was thankful for the home-cooked meal. I glanced at Lisa, who knew I wasn't feeling well. She threw a frown my way and told me from across the table, "I think I may have the same thing starting." At least I would have company I thought, grimacing.

Then two other students and a British researcher led Lisa and me up through the forest in the moonlight to our huts. The cicadas were particularly loud, and the night air was more humid than I thought possible. I took a deep breath, contemplating being alone in the dark forest all night. Walking single file, staying close to our leaders along the trail, I noticed some vines moving back and forth. At first, murmuring to each other, we assumed that the bush pigs people had described to us were active at night, or perhaps that the night breeze was moving the vines.

"Stay calm," our guides whispered.

My heart pounded in my throat, and I wished I could quicken our pace. *Where are the huts?* I wondered, nervous. I tried to catch Lisa's eye but couldn't make out her face in the shadows.

As we continued, the movements in the vines gradually reached a thrashing level, and right next to the trail loud screams, crackling branches, loud grunts, and howling began and then increased in magnitude. I panicked, unsure if I should freeze or flee, and Lisa looked at me with enormous eyes.

Suddenly, three human forms jumped out at us and burst into laughter. "Surprise!" Our fellow researchers were performing what was apparently the Gombe initiation ritual. Not very funny to us, but after recovering we could at least fake a laugh or two.

My hut was the farthest from the beach. By the time I reached it, Lisa and the others were already in theirs, slightly lower in the valley. Jim, another student, accompanied me and made sure that my candle was burning and I had a pillow and sheets for the mattress on my small wooden bed.

Until this moment on my first day at Gombe, I had felt well supported and relatively safe, but after Jim said good night and I could no longer hear his footsteps as he walked back to his own hut, I felt quite alone.

Without human voices, the sounds of crickets, bush pigs, and occasional gusts of wind grew more noticeable. I did not yet have any real sense of where I was in the forest in relation to the lake, the mountains, and the valleys. Thoughts of my noisy college dorm back in California, with hot showers and lots of friends in the hallways, peppered my deep loneliness.

As if the initiation weren't enough to frighten me that night, alone in my cabin I suddenly saw a large centipede poke up through a crack in the concrete floor. At Gombe, the researchers and field assistants were more worried about centipedes than snakes. Back home, where the tiny bendable creatures would travel through the garden on their many hair-like legs, centipedes seemed harmless, but at Gombe, they were described as big, shiny red, and highly poisonous.

As it emerged and began to crawl across the floor, I grabbed the jar I saw in the corner and went after it. The sweat from my panic began to merge with the sweat from my fever. Instead of trying to catch the creature, I gave in to my fear and tried to crush it. The edge of the jar cut the centipede in half, and to my horror, each half kept crawling. *Now I'm alone in my hut with* two *poisonous centipedes!* In my feverish state of mind, I pictured the hundreds, if not thousands, of tiny centipedes that would soon be crawling around my hut. Finally I brushed both halves out the door and put a bucket over the crack.

Perhaps the scare was good for me, as it used up my last ounce of energy. I couldn't wait to slip into bed—after checking the sheets for centipedes—and restore my energy for a big day of following the chimps.

CHAPTER TWO

SETTLING IN

A warm night breeze rushed through the screened window of my hut and blew across my face. The candle on my desk flickered and cast shadows on the letter I was writing to my parents, recounting my first day following two chimpanzees, Fifi and her son Freud, through the forest. Halfway around the world from my family, I tried to help them visualize some of the highlights of my "jungle" adventure. The wet season was ending, I explained, and the abundant plumlike parinari fruits were attracting the chimps to these trees high in the valley.

My hut, amid leafy trees, communities of chimpanzees, troops of baboons, bush pigs, and colobus monkeys in the canopies, was a twenty-minute hike from the Kasekela waterfall or the camp's main gathering place on the shore of Lake Tanganyika. I could see no signs of civilization from this spot.

For the next eight months, this would be my home. I found the hut's aluminum walls and thatched roof reassuring—they were strong enough to keep big animals out—and the cement floor was solid. The large window and its firm wire-mesh screening made a generous opening that allowed the night breeze to blow in, but also made it possible for small critters and snakes to enter. Venomous black and green mambas inhabited the Gombe forest, but given the choice of this strong metal window and the occasional slithering visitor or a flimsier fine mesh, which the more aggressive baboons could easily push through, I was content with what I had. The wasps' nest on the ceiling and the friendly geckos climbing on the walls kept me from feeling too lonely.

Pondering what more to include in the letter, I glanced through the window to catch a glimpse of the moonlight shimmering on Lake Tanganyika. The only sounds were crickets and the occasional snort of a nearby bush pig. I realized that I was just starting to feel comfortable alone at night in my small dwelling surrounded by thick vegetation and roaming forest animals.

Later that night, a huge gust blew through the window, and the aluminum door flew open, jolting me from a sound sleep. My heart pounded; I thought I had been awakened by an explosion. Then I realized it was the prelude to a fierce thunderstorm. I felt more exposed than ever to the elements. Wanting to crawl back under my covers, I instead forced myself out of bed to close the door and survey the hut for damage. The roof had good thatching, however, and the hut was tucked deep in the forest so I was at least partially protected. As I tried to settle down and get back to sleep, I told myself that if the chimps could survive high in the trees, I could survive on the ground in the aluminum-walled structure. I didn't sleep well the rest of the night.

The next morning, as bright sun shone through the forest canopy and I felt only a soft breeze on my face, I thought about the many times chimps in the wild were confronted with sudden weather changes, such as downpours, heat waves, or thunderstorms, all of which they seemed to take in stride, showing few external signs of distress. Over time, their apparent confidence with nature rubbed off on me. I didn't worry so much about getting my khaki shorts and shirt soaked when it rained, knowing the sun would appear and dry them out. I didn't get nervous about the snakes and centipedes as the chimps were never harmed by them as far as we could tell. Yet being hungry toward the end of a long day in the forest was another matter. I discovered my own food-seeking drive was just too powerful to overcome. I sometimes detoured briefly from my trekking across Kasekela Valley and ransacked the upper-camp kitchen for another peanut butter and jelly sandwich when the chimps traveled in this vicinity.

Early on, most of my days centered around getting to know the four mother-infant pairs I was assigned to study: Fifi and Freud, Melissa and Gremlin, Passion and Pom, and Nova and Skosha. All the infants in this group were between two and three years old—the perfect age for me to observe their fundamental learning and development. As I followed them through the jungle, taking copious notes and watching closely, the chimps became my "jungle professors," showing me firsthand how remarkably skilled chimp mothers are at rearing strong and adaptable offspring.

Adding to my knowledge of the larger community of fifty chimps were the "gossipy" conversations we had in the evenings at the meeting-house near the beach. The large wooden structure was our communal gathering place where the students, graduate researchers, and Jane met after swimming in the lake at the end of a long day tracking chimps. "Jomeo exerted a considerable amount of energy to capture a colobus monkey today in lower Linda Valley but immediately had it snatched away by Figan and Sherry, who took off with it before he even had a taste," Richard described to the group as we sat at the large wooden table eating dinner. We all sighed in pity for Jomeo. The result of our sharing the adventures of our particular chimps was to gain a bigger social picture of the community on any given day. Of course we also had skit night, music night, card games and intense sociopolitical discussions in the soft lounge chairs near the large windows looking out toward the lake. The field assistants were quite happy to hang out in their area of multiple smaller structures where they ate, conversed in Swahili and prayed evening prayer. The generator kept the only source of lights going in these areas but time was limited since we parted company around 9:00 P.M. to reach our huts in time for a good night's sleep. We needed to awaken before sunrise to find our chimps' nesting sites before they left to begin their daily travels. For me, that was usually the nest of Fifi and Freud. I thought about them day and night.

In a class of her own, Fifi, a young mother then at just fifteen years old, taught me the most during my stay because of her remarkable attentiveness, patience, and playfulness with her son. Her mothering skills seemed unmatchable. Though I didn't know it then, the months I would spend observing Fifi's every move as she roamed through the forest and cared for Freud, remaining patient and calm, would stay with me for the rest of my life, influencing how I treated patients and family.

The first time I saw Fifi confidently stride into our camp, Freud on her back, I was struck by the richness of her sleek black hair, which stood out against the dark green forest foliage. She certainly was not the tattered-looking, dull-brown chimp I was used to seeing in zoos back home. I recognized her immediately from the films, and I could not take my eyes off her. She was right in front of me, seeming even more humanlike than in film. Still feeling overwhelmed in general on my third day at Gombe, I smiled excitedly at Hamisi, who was sitting close by. He smiled back with a knowing understanding of what I was feeling.

Fifi stopped very close to me and rolled her son around on the ground to tickle him. I could hear Freud laughing—his rapid panting sounds accompanied by his play face were hallmarks of a chimp having a really good time. His laugh was infectious, like a human toddler's, and I found myself chuckling quietly.

I had known a lot about Fifi even before I arrived. Jane seemed to have a special fondness for her as the first female offspring of the grand matriarch Flo, and she told stories about her often. Fifi had a self-assurance like her mother's and a confidence that carried her a long way. When she became mature and was sexually receptive, she traveled several valleys south of her own community and mated with males of a different community. This behavior is crucial in widening the gene pool to produce vigorous offspring. Fifi was also very competent in her mothering and food seeking. Shortly before I arrived at Gombe I heard she single-handedly took down a young bushbuck that weighed about fifty pounds. While these antelope are commonly seen at Gombe, it's unusual for a female to hunt them on her own.

One day I wore dark Levi's while observing Fifi and Freud since my khaki shorts were all being washed. Freud noticed my different attire. As I sat with my clipboard, he slowly approached me. I caught my breath and looked around; I couldn't really move away as I was backed up against an enormous thicket. I just froze, afraid to move a muscle while Freud reached out with his little index finger. He put it on my jeans—I took care not to startle—and then he looked at me and sniffed the finger he had touched me with. I was spellbound, and impressed that Fifi was comfortable enough to allow this without snatching Freud up in her arms, although she watched steadily from fifteen feet away. A brief sadness came over me as this encounter reminded me of my close interactions with Babu. Then Freud scampered back to Fifi and ignored me for the rest of the day.

Freud had exuded a daring confidence and a competence from an early age, and I suspected then that one day he would rise to the alpha male position—which he did, eighteen years later. I'm sure this was partly due to his mother's confidence in herself and in him.

As I studied the pair, I thought of my own mother's natural confidence in being a mother and of her optimism about the world around her. Though not religious, she was spiritual and spoke of an inner strength that would guide us in life. "Just turn within to find the answer," she would say when times were rough. I related well to this form of reassurance and also to Fifi's motherly manner as she constantly made herself available to Freud for hugging and affectionate interactions during the first several years of his development. I could see this same trait in varying degrees in most of the chimp mothers I studied.

By the time I had been at Gombe for nearly a month, I could watch several of the infants playing together under their mothers' watchful gazes and tell most of the young chimps apart. Their individual faces and expressions helped differentiate them. Freud had big ears and very engaging facial expressions while he wrestled around with his peers; Gremlin's face was longer and more serious. As I watched them playing, I thought, *People will be studying these same chimps for years to come.* That thought made me happy. Jane was committed to studying successive generations in an effort to uncover deeper and more significant information about their social behavior.

My fascination with watching mother chimpanzees interact with their offspring never wavered. After I tracked the families through the forest from sunrise to sunset, my body ached with fatigue, but I never tired of witnessing the lengthy infant play sessions or the mothers' intricate nest building each night, when they carefully wove long, growing branches into cozy sleeping platforms high in the trees.

The vital bond between confident, attentive mother chimpanzees and their offspring seemed to set the stage for the offspring's successes and even survival later in life. Fifi's confidence and patience with her progeny as they learned—through observation and imitation—may have helped them secure the high-ranking positions I later learned they went on to hold as adults. Three out of Fifi's five male offspring—Freud, Frodo, and Ferdinand—achieved alpha male status. Fifi had clearly benefited from her own mother's parenting skills and likely passed that confident behavior on to her sons.

After many days of tracking and noting on-the-minute distances between infants and mothers and tape-recording descriptions of interesting chimp interactions, I would spend a day transcribing this data by hand for use in Jane's long-term study. Though I didn't realize it at the time, I was contributing to one of the longest studies of wild animals in history.

My recordings out in the field were interrupted by long periods—up to two hours—when a mother-infant pair would climb so high in a tree that my field assistant and I couldn't even see them. Whenever this happened, we'd just wait for them to come down. Far from being boring, this was my daydreaming time.

I once lost myself in watching a shockingly green, winged insect attempt to dig himself out of a spider's web. He never made it out before I had to leave, to my disappointment. Another time Hamisi laughed at me for staring into a fascinating, unusually shaped forest flower as my mind wandered. He said, "It looked like it was telling you something."

My jungle musings had been primed from an early age, when my thoughts would wander away from my third-grade teacher's lessons and on to more appealing subjects, like roller skating or climbing trees with friends. In those days, I would get called sharply back to reality by a frustrated teacher or giggling classmates, but in the forest daydreaming seemed natural and relaxing—my jungle meditation.

I also experienced some heart-stopping moments. On one occasion I noticed a large baboon named Stumptail watching me through the window of the upper-camp lunchroom as I made a sandwich for the day and packed it into my shoulder sack. I left the structure to begin following Fifi but had to bend down to tie my tennis shoe. Instantly, a golden hairy arm reached over my shoulder to grab the sandwich out of my sack. I stood up quickly and yelled, and Stumptail ran off.

I was told later by more experienced researchers that I should have surrendered the sandwich and remained still so that the baboon wouldn't startle and sink his large canines into my neck. Even with such moments of terror thrown in, I couldn't help but appreciate the raw beauty of my new surroundings. They were much more peaceful and relaxing than school, with the pressure of tests and paper-writing assignments in my premed classes.

In those first weeks at Gombe, I experienced the majesty of the African forest and its denizens. As I lay in bed at night, my tired muscles aching, I heard baboon grunts and trees swaying in the wind outside my hut. While waiting for the chimpanzees to come into view, I watched troops of graceful colobus monkeys swing across twenty-foot distances high in the canopy. One day, field assistants sent out a message Paul Revere-style, saying we should come see some birds, rarely seen at Gombe. We gathered to gaze at two majestic crowned cranes that landed and strutted across an open field near the beach. They were dazzling. Laughing to myself, I thought, *If only they knew what celebrities they are to those of us watching them.*

After what felt like a particularly long day, a number of us gathered around a campfire on the beach. For a while, I listened to people tell stories as I dug in the sand with a stick. But before long, I sunk into some sort of a trance. Perhaps it was the sheer exhaustion from scampering through thick bushes and hiking up and down hills for twelve hours to follow Fifi and Freud. Or maybe it was the hypnotic flames of the fire. Regardless, I entered an altered state. I was surrounded by chatter in English and Swahili, the sounds of crickets and conga drums, brilliant stars against a clear black sky, the bobbing yellow lanterns of fishing boats reflected on the lake, and a warm breeze. I felt at peace, relaxed, deeply connected to the never-ending cycle of the jungle. It wouldn't have surprised me to suddenly see my ancient hominid ancestors emerge from a nearby cave and join us.

I was becoming more intimately connected to my natural surroundings. I wasn't departing from reality; on the contrary, I was finding my place. The physical environment, the natural environment, was permeating me with all its history, its life, and its dangers. I had never felt so alive with ideas and imagination back home in the States. I was feeling the essence of the Gombe forest. Just as it was the living, breathing, pulsating home of the chimp families, it would become a home for me too—as these gracious primates seemed willing to share their land with a peaceful white ape visiting from California.

The Gombe Medical Clinic

When Jane asked me at the beginning of my stay at Gombe to help out in the small medical clinic at the camp originally set up by Jane's mother, I responded hesitantly. As much as I was interested in helping, I knew little about clinical medicine.

The one-room clinic had chairs, an exam table, and a supply of medications ranging from aspirin to antimalarial drugs, which had come not only from Kigoma but from the United States and England as well. In the clinic we treated field assistants, staff, and their families, some of them from Bubongo Village. One researcher, named Emilie, had training as a veterinary technician; she was the most experienced in medicine. Another, Julie, had a biology background, and she and I worked on alternating days as Emilie's assistant.

We saw people with injuries, malaria, and fevers of all kinds. We would treat them or send them to Kigoma in our park boat to be seen at the clinic there. Back pain, stomach ailments, and skin and eye infections were some of the most common conditions. We treated them as best we could. One man from a neighboring village came in with several beetles in his ear canal after sleeping on the ground. A staff member who grew up in a village close by knew a remedy for this and poured a warm, oily liquid into the man's ear, which I assumed killed the beetles, since he had no further complaints.

Some people elected to go to the local medicine man, who dispensed herbs and other natural remedies and who was accepted by many villagers as their main resource for health concerns. In rural Tanzania at that time, these natural healers seemed to be used as much as, if not more than, the government clinics.

Early in my clinic experiences at Gombe, a mother came in with her feverish baby. She spoke calmly, telling me, "He has not been eating, and he feels so hot," but her strained expression has remained in my mind to this day.

When I examined him, I was surprised by how hot his little face felt. Since we didn't have much ability to treat an infection, I taught her some basic fever care and told her that she would need to go to the hospital in Kigoma if the baby did not improve in the next twenty-four hours.

Four days later, I was eating breakfast when Emilie came in looking teary. "That baby died," she told me, and I suddenly lost my appetite. I remembered his tiny face under my hand, and his mother's worried expression. The next day, she had walked the infant to the local healer rather than go to Kigoma as we had encouraged her to do. Her belief system had likely influenced her choice. I knew, however, that the baby could have had a bacterial infection or malaria, and the hard truth is he might not have survived even if treated at the nearest hospital. Regardless, it was tremendously difficult to have to witness firsthand how vulnerable infants are to infection in tropical areas.

Hearing about the 12 percent infant mortality rate in Tanzania (compared to the 1.8 percent in the States in 1974) prior to my arrival was one thing, but witnessing it was another. There did seem to be an acceptance of this loss of young life, and people had larger families knowing that some of their offspring wouldn't survive childhood. It was a hard reckoning for me, coming from the West and trusting in medicine as the best weapon against illness. The loss of young lives is always tragic, but in Africa, it was also a part of everyday life. This new perspective was overwhelming to me. And yet that young mother's face showed me the human side of the statistics.

A few weeks later, I learned another life lesson when a physician who was also a Catholic nun visited Gombe to meet Jane. The woman was highly regarded for her missionary work in Africa. She appeared to be in her early fifties and dressed in her full habit, which protected her well from the sun but must have been uncomfortably hot to wear in the humid, eighty-degree tropics.

I was in the clinic seeing a young child with a mild fever and cough when the doctor arrived at the meetinghouse. I decided to seek her advice and walked the shy young mother and her child over for a consultation. The doctor was talking with Jane but was told why we were there. She simply nodded and kept speaking with Jane.

It seemed a bit insensitive to me for the doctor to ignore us as we waited. After about ten minutes, which seemed more like an hour, the doctor came over and introduced herself.

I told the doctor, "I'm concerned the child might be developing pneumonia."

Without even examining the little girl, she told us, "She will do fine without any medication. If not, the mother should bring her back." I couldn't understand her response but didn't feel comfortable questioning her for a more thorough answer in front of the mother. Though the doctor was calm and matter-of-fact in her diagnosis, and the mother herself seemed reassured, I remained worried.

After clinic hours, I caught up with the doctor. "I don't want to be disrespectful," I said, "but I still feel worried about that little girl I brought to you earlier. I was afraid she had pneumonia, and I was a little worried—maybe you didn't get a very good look at her."

She smiled. "Are you worried that I didn't examine her?" I nodded, feeling uncomfortable. "I'm glad you care. But don't worry. I was observing the girl out of the corner of my eye while I was talking to Jane. I didn't once in that whole time see her cough or show any sign of distress."

It was remarkable. Through her close yet unnoticeable observation, she was able to correctly assess the girl's condition and needs. The doctor's vast experience working with few resources and little technology had made her a more observant physician. She compensated for what she lacked in equipment by perfecting her observational skills in diagnosing patients. I will always remember this doctor—her calm, direct manner, and her ability to provide medical care in challenging circumstances.

The Gombe medical clinic allowed me to help meet the basic needs of a rural Tanzanian community while learning some of the fundamentals of medicine. I also hold fond memories of forming a closer connection with the field assistants' family members, most of whom I would not have met if I had not provided medical care for them along with the other villagers. And even today, when I look into the ears of my patients to check for infection, I think of the man with beetles in his ears and the oily concoction that was used to treat him. The only case that has come close to this in my medical practice was when I found a tiny sparkling Christmas ornament in the ear of a four-year-old girl. When I shined the otoscope light into her ear, I let out a big "Wow!" as I saw the glittering reflection shining back at me. I thought about using an oily concoction to get it out, but instead I called a specialist.

CHAPTER THREE

A DAY IN THE LIFE
OF FIFI AND FREUD

I spent the next few months observing mother-infant pairs. Fifi's skills as a mother made a deep impression on me and would one day help me to understand maternal nurturing among my patients. The constant physical closeness early on in a chimp's development provides deep reassurance to the growing primate. Chimp mothers also provide on-demand nursing, protection from other animals, and time for learning through the youngster's close observation of his or her mother. Young chimps gradually achieve independence from their mothers, with males starting to break away for short periods of time at around eight or nine and females at ten or eleven years of age. They reach full maturity at sixteen for males and thirteen or fourteen for females.

Melissa and Nova demonstrated effective mothering techniques as they nurtured their young single-handedly in the wild. Passion was another matter; she was typically slow to comfort and protect young Prof in dangerous situations.

Fifi grabbed my attention from the start. As she was Flo's eldest daughter, this was not surprising. Although we did not know who her father was, we could easily see that Fifi resembled her mother in her confident, affectionate nature and the way she remained nonchalant about mothering her offspring. To spend a day observing her was physically challenging at times but well worth the exhaustion at the day's end, as I wrote to my family, using field notes from a typical day in the forest:

At 5:30 A.M., half asleep, I threw on clothes, washed my face, grabbed a flashlight and tape recorder, and dashed to the *cho* (outhouse). Everything was dark and quiet, except for an occasional gust of wind.

Hot tea and toast with honey at the upper-camp kitchen helped fuel me for the day's adventure. After I met Hamisi, we set off to find the site of the chimp nest, which we had marked by placing sticks on the trail the night before. I felt increasingly awake and alert with excitement, frolicking through the forest as we neared our destination. We finally arrived at the tree where we had left the chimps the evening before. We lay back and watched their leafy mattress moving high above. Birds began to chirp, baboons grunted, and strange clicks and buzzing gradually grew in volume throughout the forest.

In the dawn light, we saw a little hairy arm stretch out of the nest. We waited quietly as the morning routine began. There was shaking and stirring, then a small spidery form crawled across branches away from the nest. Soon, a much larger form emerged and gracefully swung down to lower branches. We heard what sounded like a small waterfall as Mama Fifi took care of her morning duties, and we made sure we were not directly below.

After she'd fed on a few large milk apples, Fifi moved closer to the ground and caught the eye of Freud, her son, who had preceded her out of the nest. She waited until he approached her. He clung to her belly and enjoyed suckling milk as they descended the tree and began their travel on the ground.

Fifi signaled Freud to climb on her back, and our trip began, Freud riding jockey-style while we followed at a distance. We tromped after the pair through thick bushes and vines, over hills, and across streams. The undergrowth is especially thick this time of year, with lush grasses and small stunning red and purple forest flowers. The chimps moved through the jungle on the ground or swung through trees, giving no hint of their ultimate destination. They were usually in search of food or sometimes other chimpanzees for grooming sessions and playtime. We crawled, stooped, and jumped over branches in pursuit. The chimps, hunched down, glided with ease through the tangled foliage, while we clumsy bipeds flailed along behind. Loaded down with lunch, a tape recorder, and a clipboard, we struggled to keep up.

Fifi took a fruit break, climbing up into a gnarled, medium-sized tree to eat fiber-filled, plumlike *mbula* for an hour. This fruit grows in big clusters in trees high up in the woodlands. It has a semisweet taste, and it has just come into season. Melissa and Gremlin joined them, and while the mothers fed, Freud did acrobatics in the trees or played sociably with Gremlin on the ground. During their long play session, Fifi was constantly alert to Freud's whereabouts, and was always receptive when he returned to her for a quick suckle of milk or a reassuring embrace. I saw Freud twist the fur on Fifi's chest between his fingers, and she held him tenderly with an arm wrapped around him until he was satisfied. He then returned to the ground to resume his play with Gremlin.

After their feast, both mothers returned to the ground and groomed their infants and each other. This calm activity involved one chimp combing his or her hands through the hair of another, likely purely for bonding and relaxation. Fifi then instigated a play session with Freud; she rolled around with him among the leaves and sticks on the ground, both displaying openmouthed play faces, their laughter sounds easily heard. Gremlin was busy nursing but soon returned for more play. Both mothers then sat back and just observed the young chimps wrestling and chasing each other.

I was awestruck by Fifi's and Melissa's devotion to their infants. Their mothering seemed calm and effortless, yet it required a high level of competence as they stalked through the forest and climbed high into trees to find food, always keeping a close eye on their offspring. In addition, they nursed and provided affectionate care and protection all at the same time.

Crouching on the damp soil watching this scene, I was suddenly transported many millennia back through time, picturing early humans doing almost exactly the same things. Melissa and Fifi were like women sitting around a campsite, watching the children play amid their food-preparation duties.

I did this a lot. I would picture early humans making their way through the forest valleys in search of ripe milk apples and social interaction, like the chimps were doing before my eyes. I daydreamed about them stripping leaves from sticks to use as tools, or building places to rest with their families. I imagined daughters watching their mothers feed and care for their offspring, and pictured mothers patiently looking on while their children learned to prepare their own food.

As a youngster, Fifi had been fascinated with her younger siblings. Back at Stanford, I had been glued to *National Geographic* footage showing close-ups of Fifi's persistent attempts to take her infant brother from her mother, Flo, so she could hold and groom him. Flo usually allowed Fifi to try out her future mothering skills during these short practice sessions but was very quick to intervene when needed. In the field, I loved watching Fifi engage Freud in play, using her foot to nudge him, just as I remembered having seen Flo do with her son Flint in films.

Gremlin was one of my favorites among the young chimps. At two and a half years old, she had an intense desire to play with Freud and others her age. I often saw Freud and Gremlin playing together for hours as their mothers sat very close by, feeding or observing the play. In constant motion, the offspring chased each other, wrestled, and swung in the trees. They seemed to thoroughly enjoy being with each other.

As I watched the two interact that day, I wondered if the only reason the mothers spent time together was to let their offspring play. Fifi and Melissa might interact by grooming, but often they would just rest on the ground or feed in separate trees. I thought it made evolutionary sense in terms of allowing young chimps to learn communication skills while under their mothers' watchful eyes.

Time for Learning

When playtime was over, Fifi moved on, Freud riding on her back. It took only twenty minutes to arrive at the next food source. As they slowed down, Hamisi quietly approached me and indicated that Fifi was about to fish for termites.

One of Jane's first groundbreaking observations was the discovery that chimpanzees, like humans, not only use tools but also make them for use in collecting food and water. This literally changed the dictionary definition of "human." After Jane described the fishing for termites to Dr. Louis Leakey, he famously said in a telegram back to her: "Now we must redefine 'tool,' redefine 'man,' or accept chimpanzees as humans."

Sure enough, Fifi began searching for a long stick before I could even see the large termite mound a hundred feet ahead. "Wow," I whispered. At that moment, I truly understood how closely linked our two species are. Fifi was not just reacting to what was in front of her; she was also planning ahead—remembering that there was a good termite mound in the distance that would require using her fishing tool. Combing through a large bush, Fifi examined several branches before selecting a suitable one. Then she removed the side twigs so she had a smooth, flexible tool with which to extract the termites. She clearly understood what she needed to do to sustain both herself and her son.

When Fifi arrived at the mound, she sat down and very carefully inserted the stick into a hole she made by pushing her finger into the dirt. She held the stick very still, then gently removed it. Ten or more termites clung to it, and she nibbled them right off the stick. It might sound very basic, but most human researchers who attempt this skill initially rate quite low on the achievement scale. The problems seem to arise with human clumsiness and timing during the process of trying to get the critters from mound to mouth.

I watched Freud study his mother at the termite mound. Fifi allowed Freud to practice the basics of termite fishing by poking a stick in the dirt inches from where she was working. Though ineffective at first, by age five, Freud would become very efficient in collecting this high-protein food. Jane's historic discoveries were more than just exciting moments that

I had the good fortune to witness. I could also see how crucial it was for Fifi to be patient with Freud. She needed to allow him the time and space to try to copy her termiting technique. His survival would depend on it.

Not until later in my life would I understand how Fifi's example of patience and steadfastness could apply to human teaching and learning. Though I didn't yet have the life experience to directly apply what I was learning, the image of Freud closely watching Fifi termite was etched into my memory and emerged later as guidance for fatherhood and doctoring. "Be patient, be patient, be patient," became my mantra during my sons' terrible twos, as they boldly explored their surroundings and exhibited their emotions without restraint.

I did bow to social pressures at times. In one instance, when a neighbor was over and Tommy began banging and throwing pie pans across the kitchen floor, I asked myself: *Is my two-year-old's throwing pie pans actually dangerous, or is the perceived problem more the frown on the face of my visiting neighbor?* I took the pans away from Tommy—but wasn't entirely comfortable with my decision. I felt sure Fifi would have let him carry on exploring his environment.

"Be patient" was my mantra again during my sons' teenage years, when they exerted their need for independence. Structure and boundaries were necessary, of course, as they are in any child-rearing, but for me the most difficult task was holding back. During their teens, I told myself, "Just be present," when my impulse was to be more forceful. Their aloof manner, their ignoring certain requests we made, and the remarks intended to make us feel less than smart were all there.

But I would recall Fifi's unwavering focus on Freud and her responsiveness to his needs through the simple act of paying attention. I learned later, after I left Gombe, that when Freud became an adolescent, Fifi continued to be tolerant of his intimidating displays and daring interactions with the adult males while still being present for grooming and brief hugs when he needed them. Fifi taught me that modeling calm and being observant were more effective in the long run than raising my voice (although I certainly gave in to loudness on occasion).

Our journey through the forest continued. I was thankful that during most of the day, I was shaded by the forest canopy and that the afternoon breeze cooled me down. As Hamisi and I forged through thick brush and leaped over streams, I wondered if the shorter stature of the chimps made their traveling easier, as they effortlessly weaved through the plant life on the ground and glided across branches in the trees.

There was no naptime for Freud unless he dozed off for a few minutes while riding on Fifi's back as she traversed the forest ground. When Fifi stopped to rest and Freud nursed, he did seem to nod off now and again. Chimps in the wild aren't weaned until four or five years of age, so the period of contact with the mother is prolonged, allowing intricate learning to occur as well as guaranteed nutrition. The lengthy mother-infant intimacy period is much like that of our own species, and shows how important early bonding is to any infant's development. The constant natural physical contact benefited Fifi as well. It clearly calmed and energized both of them and continually reinforced their mother-son bond in a way that was completely integrated into their life together. After a thunderous display by Figan near the waterfall, for example, Freud scampered into Fifi's arms as she reached for him. The two clung to each other until Figan's aggressive behavior ended. Freud knew where he could find safety, and the two were tethered by this bond in the unpredictable forest.

Fifi needed no designated "quality time" with Freud—all their time together was seamlessly connected.

In an excerpt from a field report I wrote at the end of my stay, I described some of the unique characteristics that set Fifi and Freud apart from the other mothers and infants at Gombe:

> One of the most dynamic and self-assured teams at Gombe, Fifi and Freud interact with each other in play more than any of the other mother-infant pairs. Fifi is very attentive to Freud, even though this may involve just a glance in his direction when she is stretched out on her back resting and Freud is engaged in social playing several meters away. While she feeds, there are long periods of time in which Fifi pays little or no attention to Freud, but a significant amount of the remaining time is spent in very positive interactions such as play tickling, wrestling, grooming, or watching Freud's social play.
>
> Though Freud appears at times to be completely uncon-cerned about Fifi's whereabouts when he's involved in play sessions, he's usually very attentive and responsive to her move-ments. There seems to be very little tension between them. They've worked out a remarkable communication system by which Fifi uses play to initiate travel. She will approach Freud, play bite him or just slap at him in play, then begin travel. Freud responds quite well to the signal and is redirected to follow her.
>
> One characteristic feature of Freud is his ability to engage in vigorous locomotor play and pirouette during travel. When Fifi would get several meters ahead, Freud would quickly scamper after her, sometimes pirouetting so energetically that he would fly off the trail and end up in a bush. He would then run to catch up with his mother.
>
> Perhaps it was fun for Fifi to have such an acrobatic son.

Fifi and Freud made their way up the valley. In late afternoon, I watched as Fifi slowed suddenly to listen to the pant-hoot of her brother Figan across the valley. She clearly recognized his voice, and seemed to recognize the context as well.

"Maybe Figan found some ripening fruit," Hamisi said, trying to interpret their communication.

Fifi crossed the valley with Freud on her back until she found Figan, who was feasting on milk apples.

"You were right!" I said to Hamisi with surprise. Though I knew that subtly different pant-hoots had different meanings, I couldn't yet tell them apart like Hamisi and the chimpanzees could. During Fifi's formative years, she'd had ample opportunity to learn the group's signals. This was another reason the chimps' close relationship with their mothers was so important. The mothers guided their young in learning these all-important skills. Just as is the case for humans, strong, instinctive communication skills are critical for chimpanzee development, survival, and social success.

Evolutionary Perspective

Nova and Skosha must have heard the call too, because the mother-infant pair showed up to join the gathering group of about fifteen chimps. Things were quite peaceful as they filled their stomachs with milk apples and socialized. Socializing is as important to chimps as it is to humans. It's a chance to release energy, bond, and communicate—in short, to play and have fun. Watching the way the chimps socialized among themselves—young chimps and mothers, along with other members of their community—reminded me of large family gatherings like cookouts or playing in the park. This aspect of chimp life is crucial to the overall well-being of both the individual chimps and the group as a whole.

Nova and two-year-old Skosha were soon climbing high up in a leafy parinari tree. I stopped watching them for a moment to focus on brushing a group of lethargic biting flies off my arm; when I looked up, I saw Skosha swinging from its branches, performing her usual acrobatics. Then, a black furry ball—Skosha—suddenly dropped seventy feet and landed in the thick brush below. I sat there, stunned, thinking she would not have survived the fall.

Nova rapidly descended to where Skosha lay very still and began to inspect every part of her body. She diligently licked her daughter's wounds, groomed her, and then sat very close until Skosha was able to move on her own. It took a couple of hours, but Nova never moved from Skosha's side. Eventually, Skosha was back to swinging high in the trees, her mother staying close by.

There was something very primal about Nova's response to her infant's fall. She demonstrated how her focused attention after trauma comforted and reassured Skosha, as she checked for serious wounds. I later witnessed skilled human parents respond in similar fashion after illness or trauma to their children. On the other hand, I've seen overprotective parents worry so much that their children became anxious too.

As a father, I could feel anxious about trauma or illness in my children, but I didn't want to convey my worry to them. I wanted them to have confidence that nature would heal their condition, and if not, that help would be found. I appreciated my mother, who had always remained calm in the face of our broken arms and other childhood injuries. I cringe when I see parents too worked up over simple colds or muscle strains in their children; I cringe too when they show little or no attention. The chimps demonstrate a balance: they attend calmly and thoroughly and then move on.

The group continued to feed on milk apples and communicate through vocalizations, grooming each other, nursing, sitting side by side, and staking out the right trees for their evening's nest building. Fifi's attention was solely on Freud as they played, and she remained in close proximity. Nova continued her vigilant care of Skosha, and Melissa gently groomed Gremlin following a big play bout with Freud. Several adult males sat quietly nearby, grooming with adult females or resting alone.

I let myself relax and took a mental step back to view the entire scene in front of me from an evolutionary perspective. The chimps were thriving in this remote forest because they were doing the right things to survive. Over millions of years and significant changes in their environment, their genetics and their passed-down culture have kept them alive and reproducing in their Tanzanian habitat. Thinking about the footage I had watched back at Stanford, I wondered how many of Fifi's mothering skills she had learned from Flo and how many were programmed from birth. It's the age-old question: How much does environment shape behavior, and how much is inherited through our genes? To add to the mystery and intrigue, scientists have recently discovered that active changes can occur in our genes during our lifetimes in response to environmental events. These altered genes can then be passed on to the next generation. Whatever the combination, I could see that the chimps' mothering techniques proved highly successful, and we can surmise that the long period of dependency allowed for plenty of nurturing.

The cultural differences among various communities of wild chimpanzees strongly support the idea that learning by observation plays an important role in their survival. One of the postdoctoral researchers spoke about this at dinner one night, saying, "Chimps in other areas of Africa learn from their elders how to break open nuts by hammering them with a rock. Since these nuts don't grow at Gombe, 'Jane's' chimps don't demonstrate this skill; instead they've learned the skill of extracting termites from their mounds. Learned ages ago, the skills of termite fishing and nut cracking were and still are passed from one generation to the next by observational learning." I was fascinated by the discussion.

In the larger evolutionary scheme, a young chimp's long period of learning from his or her mother might also be helpful to chimp communities adapting to an environment that has changed over the centuries. As new environments emerge, creating different food sources, the chimps gradually learn to exploit their surroundings for the nutrients required to survive, and then they pass those new skills on to their offspring.

Whatever instincts both humans and chimps possess, it appears that new mothers of both species can benefit by watching competent mothers in action. When chimps are raised isolated in cages for research, they have difficulties with mothering, at least initially. In "Primiparous Chimpanzee Mothers," author Mollie A. Bloomsmith discusses how female chimps raised apart from other chimps were totally incompetent with their first offspring. Even those raised in a socially and environmentally complex nursery without their mothers often lacked these all-important mothering skills. Only one-third were capable mothers.

Eight years later, I was able to use what I learned from Fifi and Freud when I was shut into a small examination room with a fussing two-month-old and her mother. With a grim expression and dark circles under her eyes, the young woman said, "This is basically what Jenny's like all the time. I don't know what to do. The house is a mess, and I can barely get anything—" Her voice cracked, and the baby started crying in earnest. "I can't get anything done. I don't—" The woman started to cry too.

I felt terrible. I patted her arm, and reached for the tissue box. I had no personal experience with colic at that point, but I knew that it could feel like a never-ending problem to an exhausted parent. I also needed to make sure she did not have postpartum depression in addition to her more obvious struggles.

"I'm sorry," the mother said.

I opened my mouth to speak but suddenly had a flashback to Gombe. I remembered Fifi going about her day, performing her sometimes-challenging tasks with Freud clinging to her for reassurance.

"Do you have a baby carrier you can wear on the front?" I asked the mother as she dabbed at her eyes. She nodded. "I want you to try something. Put Jenny into the carrier and let her ride close to you while you work around the house."

The woman looked interested but a bit skeptical, though she said she would try it.

When I next saw the pair, several weeks later, they both appeared more cheerful. "Things aren't perfect of course," the mother admitted, "but Jenny seems to be doing a lot better—and me too!"

Having witnessed the importance of close physical contact between primate mother and infant so clearly exemplified by Fifi and Freud, I had been able to help the human pair. The close contact, the movement, and the jiggling that can help expel a baby's pent-up gas helped the mother get through her exasperating time. Since I had no children of my own yet, I had only my chimp observations from which to formulate practical parenting advice. Many new moms get helpful hints from relatives and peers, but I felt the need to contribute to the parenting formula when asked. Despite having some very progressive parents in my practice, I never revealed the true source of my practical recommendations.

Most of the mother chimps I observed at Gombe demonstrated a high degree of attentiveness and consistency with their offspring, but many researchers were critical of Passion. She traveled more on her own, away from the other mothers, and seemed to ignore—or at least respond slowly to—the needs of her daughter, Pom. Even the way Passion handled her two-year-old son, Prof, seemed rough. A researcher told me that Passion could clearly have used some "peer training."

Later, the mother-daughter pair shocked Jane and others—for a period of time Passion and Pom snatched baby chimps away from their mothers and killed them. There was no clear explanation as to why they cannibalized at least three and probably more infants in the Kasekela community. One theory was that it was purely for ease of getting meat, since on one occasion, Passion even embraced Melissa after she had just killed Melissa's infant, as if to convey that she hadn't committed this act out of anger toward Melissa. Adult males occasionally cannibalize infants, but they take them from females of different communities—strangers. The cannibalistic behavior of the mother-daughter pair finally stopped after Passion and Pom both gave birth. Fifi was the only mother who had a surviving infant during those three years. I wondered if the seemingly unnatural behavior of Passion and Pom was at all linked with the lack of maternal warmth or if there was some other innate cause.

We never intervened if a chimp was in trouble. We wanted only to observe and record their natural behavior in the wild. However, Jane made an exception to this rule early on when polio spread from a human village to the chimps causing limb paralysis and even infant deaths. Since it had started with humans entering the chimps' territory, she felt it made sense to insert polio vaccine into bananas that the chimps ate so their community wouldn't die off from a human-transmitted disease. With this exception, we otherwise witnessed what our early ancestors likely faced in terms of the challenges to stay alive and reproduce.

At Gombe, I pondered *human* behavior in relation to the nature-versus-nurture theme. I couldn't help but look at Jane Goodall's own family history. Jane's exceptional endurance, her focus, and perhaps her willingness to take risks might be traced to the genetics of her father, a successful race car driver. Since Jane's father wasn't around during much of Jane's childhood, Jane's mother, along with a nanny, played the major role in nurturing Jane during her childhood. Jane and her mother had in common their empathetic, strong interpersonal skills and sensitivity.

I even observed Jane using techniques that successful mother chimpanzees use. Just as Fifi and Flo mothered their young, Jane herself gently guided her then-seven-year-old son, "Grub," in his exploration of the natural surroundings at Gombe. On a warm afternoon near the beach, I saw Jane place some dry pods from a bush on a rock near the beach as

Grub watched intently. As the pods heated up in the sun, they would burst open, firing the seeds into the air as nature's way of dispersing them. She asked Grub why he thought it was important for this to happen and I distinctly remember his thinking carefully about an answer.

When I was in Gombe, we didn't know who the chimps' fathers were, though DNA testing of stool samples now makes it possible. We knew that the mothers provided most of the nurturing, regardless. Perhaps some basic mothering instincts are programmed from birth, but many of those skills are likely refined from young chimps observing their own mothers.

Three-year-old Gremlin watched Melissa closely and for long periods. She was clearly learning from her mother's example. I even saw Gremlin gently holding on to her mother's arm and staring intently as she learned to termite and build nests.

At about age nine or ten, a male chimp's adolescent hormones surge, and his brain changes as well. These changes signal him to gradually leave his mother for longer and longer periods. He begins to hang out with the adult males or on his own. Just as we humans have a difficult time navigating adolescence, so do chimps. An adolescent male will experience some tense times observing the powerful adult males from a peripheral location. He will then return to his mother for affection and reassurance. While he begins to carefully interact with the adult males, he will still often retreat to the sidelines for safety. Just like teenage boys, the nine- or ten-year-old chimps can look like and seem like small adults, but they're still preadults, their behaviors not fully formed. These preadult males may also start to occasionally behave aggressively toward adult females and, eventually, low-ranking males. Chimp males achieve full maturity at about sixteen or seventeen.

Females tend to separate a bit later. They often closely observe their own mothers raising a younger sibling. They begin to travel on their own for several hours at a time, but seek—and receive—intense hugging and grooming from their mothers when they reunite. Jane described Fifi at age nine getting separated from Flo during a rainstorm. The thunder kept Flo from hearing Fifi's whimpers and screams while they remained separated for over an hour. After the storm, Fifi climbed high up in a palm tree to scan the forest and listen for her mother's calls. She soon spotted Flo and scampered down the tree. When the pair finally reunited, there was prolonged hugging and grooming between them—and Fifi was reassured and made to feel safe once more.

In chimp society, families, including siblings and especially mothers and offspring, remain close throughout life. The numerous years that young chimps spend sheltered in close contact with their mothers create a bond that remains significant through old age. During adolescence, the closeness wanes, but even adult male and female offspring are known, as Jane notes in her talks, to spend more time with their mothers than with other adults in the community. Moreover, older mother chimps in threatening situations are often defended by their more able-bodied offspring. The pull of that maternal relationship remains strong even after the offspring leave and become independent.

The sun was approaching the horizon when Hamisi and I exchanged glances, knowing our workday was coming to an end. We watched as Fifi swiftly climbed forty feet into a nearby leafy milk-apple tree to search for a spot to build her nest. In a neighboring tree, Nova simply remade an old nest that had been previously built by another chimp. Using fresh leaves and small, pliable branches, she added more padding.

Fifi, however, managed to go far out from the trunk of her tree to a very bouncy location. Using all fours, she brought many flexible branches together by bending them into a platform. She broke off smaller branches and wove them in, then used her body to mash down the still-living long branches to continue the construction. After about five minutes, she rolled onto her back and sank into the nest. Freud immediately jumped on top of her and cuddled close. With the two comfortable and secure, no sounds other than breathing would come from the nest until morning, so as the sun began to set, Hamisi and I journeyed back to camp.

CHAPTER FOUR

OF CHIMPS AND MEN

While Fifi and other chimp mothers at Gombe provided examples of the importance of crucial nurturing behaviors in the wild, the males provided another key piece of the chimps' story. Figan, Satan, Faben, and other adult males demonstrated raw male behavior from an evolutionary perspective—behaviors critical to the survival of this magnificent community. Following the males through the forest was definitely more anxiety-provoking for me than observing mother-infant interactions. A sudden display of aggression or an organized hunt and killing of a colobus monkey by a highly energized adult male could occur without warning.

I got to know then-twenty-one-year-old Figan better during a big storm. In Africa, storms have a sound and a feel all their own. I was at the waterfall with my field assistant Rugema watching the chimps. The plunging water hitting the rocks crashed behind me. Rain pelted down and winds howled. Just then, Figan managed to kick it up another notch. It may have been the sound of the waterfall itself that stimulated him to perform. The relentless sound of the falling water seems to stir a primitive instinct in the chimps. In humans, noise also can trigger a fight-or-flight response. It registers in our psyche as "danger."

As Figan began his vibrant pant-hoot, I noticed other chimps moving out of his way. Rugema and I also moved back to distance ourselves from Figan's dramatic display of aggression. "Stay here," Rugema instructed. "Be like a tree and don't move." I continued watching, frozen to the spot where I stood.

With erect hair and a grimace, Figan grabbed low-hanging branches and swung powerfully out over the water. He landed on solid ground and continued to charge around, throw palm fronds, thump on tree trunks, and make increasingly louder pant-hoots that echoed down the valley. Melissa crouched nearby with Gremlin clinging to her torso. With a dropped jaw, I felt as if I were watching an action-figure movie, but it was more likely that I was witnessing a primal scene of an alpha male showing the community that he had not lost any of his power or confidence.

I could clearly understand why such high-energy displays evolved. Although the alpha male usually directs his displays toward the members of his community as a way of showing off his dominance, an aggressive swagger and dramatic display such as Figan's would instill fear in even the most dangerous of other forest creatures. It certainly made me nervous when I was close by! I saw adult males constantly practice these impressive and hair-raising actions—charge down hillsides, hurl large branches, and often stand bipedal with hair erect to make them look taller.

Just as is the case in human families, relationships among chimp siblings could be charged and complicated. Figan was Faben's younger brother. Faben often acted as a kind of foil for Figan, enhancing his displays of power. A few months before I arrived, Faben had helped Figan attain alpha male status by supporting him in fights. Figan and Faben worked on a coordinated attack on Evered, who was the main obstacle in Figan becoming the community's alpha male. It worked, as Evered became submissive to Figan following this "ambush." Fifi's brothers had formed a stable coalition, and this tight, protective bond helped Figan maintain his alpha status for several years.

I was amazed by such complex relationships and alliances in the chimp community. They seemed similar to human coalitions formed for political and social purposes. The genes deeply embedded in our primate cousins related to aggressive tendencies are likely similar to our own, dating back to primitive times when they may have been more useful than in today's modern world. Getting to know the chimps was like getting to know distant family members—it was a cross-species reunion of sorts.

Figan's power play reminded me of the challenge we humans face in maintaining peaceful coexistence in the modern world. One of my professors, Dr. David Hamburg, has conducted fascinating research on primate aggression, which he believes helps us better understand such behavior in humans. His work has highly relevant real-world applications. He has served as a consultant to several US presidents, advising them on counterterrorism strategies. Dr. Hamburg successfully applied his knowledge when negotiating with Congolese rebels who kidnapped four students from Gombe a year after I left. He described moments of progress during a meeting with the kidnappers. They were making demands and then suddenly broke into a violent rage. During those moments, there was grave uncertainty about the safe return of the students. Drawing on his understanding of emotional flare-ups in non-human primates, Dr. Hamburg remained calm and focused and was able to keep the dialogue going. I believe Hamburg's comprehension of aggression and how it had evolved in humans helped him secure the students' release after a month of agonizing negotiation.

Mike, a chimpanzee from Jane's early days at Gombe, rose to the top of the male hierarchy by using brainpower more than brawn. He learned that hurling empty metal kerosene cans he had found in camp down a hillside produced a terrifying sound that intimidated other males. Jane's former husband, the late wildlife filmmaker and photographer, Hugo van Lawick, caught some wonderful images of this behavior in the *National Geographic* film *Miss Goodall and the Wild Chimpanzees.* The footage shows Humphrey and some of the other males sitting peacefully on a hillside, and then suddenly leaping up and running helter-skelter when Mike's loud banging starts up. Mike was able to maintain alpha male status for several years using his ingenuity with such props. This jungle scene evokes in the viewer a strong identification with the chimps, as we can see ourselves or other humans using the same creative skills to get our needs met. It can be startling—and illuminating—to see how closely we resemble our chimp cousins.

My second physical encounter with a chimp was with Figan, who was charging down a hillside. I didn't have time to move out of his way. Terrified, I wedged up against a dense thicket, but I couldn't escape his charge. I was frozen in terror. The powerful eighty-five-pound Figan was four times stronger than a human his same weight. With his hair erect and his face as fierce as any warrior's, he dragged a large palm frond in his right hand; his left was free to swat me as he thundered by. It was more of a firm pat on my thigh than a blow, but I was so relieved that he didn't run me over. Perhaps because of that, I developed a fondness for Figan, knowing he could have ripped me apart but didn't.

Male chimps learn display behaviors early in life by watching other males. Freud was only two when I saw him try to imitate seventeen-year-old Satan, who had just completed his maturation into adulthood. As Satan came thundering through our camp, hurling palm fronds and beating against a tree, chimps and researchers immediately moved aside. Young Freud watched quietly from behind a tree with Fifi. When all was clear and Satan long gone, Freud practiced his own display by trying to imitate Satan. With one-tenth the force, he gave a few soft pant-hoots, tossed some twigs in the air, wobbled as he stomped on the ground several times, and then quietly sat down. I think I was the only primate watching this attempt at a display, and I wanted to pretend for this earnest young chimp that I was frightened.

Actual physical fighting between males does take place, just as it can between young men or women. Male chimps fight to establish a social order, with a dominant male maintaining his position for up to several years before being dethroned. About half of male-male aggression is due to competition over status. Serious harm rarely occurs. The resultant pecking order serves to keep fights from occurring more often—usually around food sources and during mating. The hierarchy assures that each chimp knows his place. The alpha male may mate with a receptive female without having to fight off other males. Females seem to have a less defined hierarchy, but their strength and ability to defend themselves—even with an infant clinging to them—is impressive. There is only occasional fighting among female chimpanzees, sometimes over meat sharing or with immigrant females from a different community.

Still, a less aggressive male might consort with a female who is in estrus (sexually receptive). We jokingly called this "going on safari." A quiet courtship develops, and the male seems to seduce the female (or vice versa) into traveling miles away from the community. After I departed, Caroline Tutin described Satan using sophisticated social skills, patience, and strategy to carefully lead Miff away from the group. The female's mating exclusively with the lower-ranking male during that weeklong period widens the community's gene pool. This non–alpha male accomplishment may also help to select for other inheritable traits, such as strategizing.

Apart from the sudden bursts of adrenaline that can trigger fight-or-flight with the enemy—snakes, buffalo, leopards, or other chimps from different communities—there's a lot of downtime for the Gombe chimpanzees. They often rest by a stream or feed in a newly ripened fig tree for an hour or two. Even Figan did. As the alpha, he in particular needed to be able to pounce at any time. He could restore his calm almost instantly, however, by hugging another chimp for a few seconds or by prolonged grooming—in his case, usually receiving it. The social activity of grooming involves gentle touching and caressing. A chimp fingers through the hair of another chimp, appearing to search for bugs or burrs, though there are usually none to be found. It's like back scratching and can occur in big groups or one-on-one. Groups of two to ten or more adult chimpanzees or a mother-infant pair will sit quietly and pick through each other's hair in a very affectionate manner. In addition to being very relaxing and restorative, it appears to be a form of communication that strengthens the bonds of family and clan.

Physiologically, such calming behaviors provide a better balance for primates' long-term health than does a constant hyped-up state. Studies in humans and other primates show that constantly circulating high levels of adrenaline and cortisol can weaken the immune system and cause long-term health problems such as fatigue, weight gain, digestive disorders, and depression. We're only just beginning to understand the physiological effects of long-term stress on humans, but we know that a continuous state of stress is unhealthy.

Often in my practice I will see patients such as Maryann, who came to me with intestinal problems referred to as irritable bowel syndrome. In her case, we could trace the onset back to when her teenage son began acting out and having major problems in school. Although she assumed the cause of her constant abdominal discomfort was a bacterial infection, it improved as she sought medical help and psychological help to calm her constant worries and stress about her teenager.

Chimps in the wild appear to have developed other behaviors to alleviate stress as well. As a higher-ranking male traveling in the forest meets other males or females of the community, his hair becomes erect, sometimes triggering other males to display erect hair. As the highest-ranking male tenses with aggressive posturing, the others usually crouch and make soft grunting sounds of submission. This is followed by reassuring hugging, kissing, and eventual grooming, which quiet the heightened energy and soothe the group.

One day at Gombe I saw Fifi do something I hadn't seen before: as her brother Figan was revving up for a display, rising onto his hind legs and with hair erect, Fifi reached from behind him and gently held his testicles with one hand. Figan immediately sat down, his pant-hoot softened, and he began to calmly groom himself. My field assistant, Yasini, placed a hand over his mouth to mute his chuckling at this scene. Just as grooming is reassuring and calming, Fifi's gentle touch to Figan's genitals pacified him and stopped an aggressive, tense display in its tracks.

This sexual calming strategy reminded me of bonobos, primate cousins to both chimps and humans. Formerly called pygmy chimpanzees, bonobos live in the south-central Democratic Republic of the Congo. They're slightly smaller than chimpanzees and have less bulky bodies, more like ours. Bonobos use sex and sensuality to defuse aggression. They literally make love more than war. They use sex in a diverse way that extends far beyond mating. This includes males mounting males to dampen aggression, as well as two females rubbing together for friendly bonding. Neither of the latter behaviors results in orgasm. I recently became interested in this species when reading *Bonobo Handshake*. Author Vanessa Woods describes her growing love for these magnificent primates living in a sanctuary in Congo's capital. Perhaps because of their specific rain forest habitat in the Congo Basin over millennia, they do not need the forceful protection and dominant displays of aggression required of the Gombe chimps to survive. Now their survival depends on an end to poaching and land destruction of their home in the rain forest.

Unlike chimpanzees, bonobos have a matriarchal society. But Jane Goodall has been quick to point out that even in the well-documented aggressive, patriarchal chimpanzee society, there are also examples of nurturing behavior by males. Gombe chimps Mel and Darbee were orphaned at age three-and-a-half, when their mothers died during a pneumonia epidemic. Each was subsequently adopted and cared for by young males in the community. Adolescent Spindle and young adult Beethoven, both of whom had lost their mothers, nurtured Mel and Darbee by sharing sleeping nests and attending to their daily needs. Interestingly, recent data on paternity (using DNA from feces samples) shows that Beethoven, who cared for Darbee, was actually her biological father. Although we don't know enough about chimp DNA and hereditary behavior to draw hard conclusions about how fathers might identify with their offspring, it's a fascinating piece of information.

Another example of male nurturing is documented in *Chimpanzee,* filmed in an African rainforest. This film highlights the true story of Freddy, an alpha male chimp who adopted and cared for Oscar, whose mother had died. Even the most powerful male chimpanzees are apparently capable of nurturing young chimps and providing care for them.

At Gombe, I was fortunate to be in the company of Richard Wrangham, who studied male behavior in chimpanzees and then later in his life became a MacArthur fellow and professor of human evolutionary biology at Harvard. Richard coauthored with Jane's biographer, Dale Peterson, a moving and highly relevant book, *Demonic Males: Apes and the Origins of Human Violence.* He described how the males in our society have inherited aggression and violence from our chimp-like ancestors five million years ago. Although these genes are no longer needed for human survival today, their presence can cause major upheavals in male-dominated societies.

Modern humans might benefit from organizing our societies more like that of our bonobo relatives, who project more sexual equality and less aggression. Our bigger brain size and moral code become crucial for our future survival and must be used wisely as they compensate for our "demonic" tendencies. From my own observations of male chimpanzee behavior in the wild and from Richard's in-depth analysis, I began to understand the reasons for aggression and competition in the chimps' world. This in turn helped me better understand these behaviors in my own species.

CHAPTER FIVE

MADAME BEE HOLDS HER OWN

About three months after my arrival, I woke up before dawn one morning, threw on khaki shorts and shirt, tested my tape recorder, and headed to the beach camp. After tea and toast, I hiked a short distance to the field assistants' quarters to meet Esilom, my assistant for the day. With a full moon to light our way, we strolled along the beach in companionable silence, passing three valleys before arriving at Kahama Valley, where the "southern" community of chimps lives.

Earlier in the week, one of the field assistants, Rugema, had announced to Jane and a few of us students, "I spotted Madame Bee and her new infant traveling in Kahama Valley, very close to our southern border."

Madame Bee was a mature chimp who was rarely seen in Kasekela Valley, where Fifi and her community roamed. Madame Bee was part of the southern Kahama community that had split off from the larger northern Kasekela group a few years earlier.

When Jane heard the news of the birth, her face lit up. Then she paused, turned to me, and asked, "John, how would you like to travel to the Kahama community to see how Madame Bee is doing with her new infant?"

Although this ambitious assignment surprised me because I was still a newcomer, of course I quickly agreed. Later that night I sarcastically quipped to my fellow researcher Caroline, "I did have to check my busy schedule before I gave Jane the OK." This was definitely one of my more exciting fieldwork assignments, and anticipating my journey south to unfamiliar territory added suspense.

Jane's attentiveness to my work was encouraging and exciting. Though she didn't usually go with me into the field, we often talked in the evening about my observations. Rather than be overprotective of the process, Jane would turn the student researchers loose in the forest, trusting them to give her accurate documentation of the chimps they studied. I felt honored that she was allowing me to go follow Madame Bee.

Our early-morning walk along the beach was pastoral. The full moon still lingered, and all was quiet except for the lapping of waves on the lakeshore. We passed several fishermen's small grass huts, which were deserted, as the fishermen never worked on a full moon. The men would sail out at night to catch small fish called *degas* with handmade nets, using lanterns to attract the fish to their boats; a bright moon would interfere with this technique. On this day, the men were with their families in the villages just over the mountains.

Esilom and I finally met up with Rafael, a guide who knew the southern community of chimps. We headed into the valley where the Bee family nested, looking for Madame Bee, who traveled with her infant, Little Bee, and her nine-year-old daughter, Honey Bee. It was important to find the chimps' nests before they awoke and began to travel.

At our destination, I gazed up at the nest at sunrise. I felt a flash of insecurity, realizing that I was in an unfamiliar valley with a community of chimps I didn't know. I felt a bit homesick for the chimps I'd grown to know in Kasekela Valley. Since researchers didn't follow these southern chimps much, I was uncertain how they'd react to researchers trekking after them. Would they be frightened and flee, or become aggressive and possibly attack us?

As the sky lightened, a female chimp emerged from the nest, her infant clinging to her belly. "That is Madame Bee," Rafael whispered. Bee's older daughter, Honey Bee, had slept in her own nest, and she soon joined them. I felt a twist in my stomach when I saw that Madame Bee cradled her infant with just one arm while the other arm hung useless at her side. I made eye contact with Esilom; he nodded slightly. Madame Bee's arm had been paralyzed during the polio epidemic that swept through the chimp community in 1966.

The illness most likely spread from people in a nearby village that had experienced an outbreak of polio just prior to its appearance in the chimps at Gombe. According to information released in 2005 by the National Human Genome Research Institute (NHGRI), human DNA is 96 percent identical to chimpanzee DNA, so it's not surprising that chimps can catch many illnesses from us—everything from common colds to polio, measles, MRSA (methicillin-resistant Staphylococcus aureus), and other bacterial, viral, and parasitic infections.

As I observed her, however, I was happy to see that Madame Bee could swing through the trees with one functional arm, moving slowly and carefully, as Little Bee clung to her belly. I was impressed with how this mother chimp had adapted to her life. It was an inspirational sight—a chimp whose disability did not prevent her from doing her job as a mother. Soon after birth, infant chimps are able to cling to their mother's torso hair and remain safely attached to her belly as she travels through the forest. This is certainly good for survival, as it frees up both of the mother's hands for traveling and the arduous task of building a nest high in the trees while her infant remains securely tucked underneath her. Even so, I couldn't imagine how Madame Bee built a nest with one arm while caring for Little Bee at the same time. As Madame Bee fed on milk apples, Little Bee nursed and Honey Bee explored the ground below. I sat, quietly taking in the views of enormous leafy trees and vines and listening to a multitude of animal sounds.

I was jolted out of my reverie when loud pant-hoots and screams suddenly started coming from higher up the valley. The Kahama males, who had nested a hundred yards away, had awakened and were journeying through the forest.

The noise grew even louder, and I gripped my notebook. Branches thrashed, and then, just fifty feet away, three male chimps charged straight at us. I realized I didn't know which males were dominant in this community, and my panic intensified. I was crouched and ready to head in the opposite direction for safety, but Esilom moved up next to me and said softly, "We're OK here."

The commotion was residual excitement over a colobus monkey kill the day before. Each of the three males had caught a colobus, and one of them, Godi, was displaying his catch by waving the carcass. I knew the difference between the play face—open mouth and relaxed lips—that chimps make when playing with each other, and the aggressive grimace, teeth flashing, made during charges. What I saw wasn't play faces.

At one point I thought things were calming, but I was wrong. Godi's display was still building as he flung branches and stampeded across the ground, screaming loud pant-hoots, his hair fully erect. Then there was a few minutes' respite as he resumed eating more of the colobus carcass. Other chimps sat nearby, begging for scraps of meat.

Just as I was hoping the show was over, the lead male, Charlie, arrived and the drama escalated. Females screamed. The earth shook. My two field assistants and I tried to shrink into the background so we wouldn't become targets. There were only seven adult males, but it seemed like twenty.

As Charlie approached, all the males began to display again. Then Godi and another male, Sniff, suddenly turned on Madame Bee, who had retreated up a tree. To my shock, the large chimps hit her and eventually pushed her and Little Bee out of the tree. There was no clear reason for this attack other than to show off their power. I grew angry watching poor Madame Bee scream and try to protect herself and Little Bee as they crept along the ground. They escaped from the commotion a bit shaken, but uninjured. Madame Bee, Little Bee, and Honey Bee finally stopped screaming but stayed very close to one another while grooming, eating fruit, and traveling after that.

Meanwhile, the males and a few females in the group scaled a huge cliff by grabbing hold of dangling vines to hoist themselves up. Esilom and I stayed with the Bee family as they traveled slowly on level terrain, with Rafael helping us navigate the unfamiliar valley.

I was quite shaken by the male violence. Watching a disabled mother chimp assaulted with such force was deeply upsetting. Perhaps the reason for this attack was hardwired into these primates' survival instincts. I thought about the inescapable fact of human aggression. Bullying came to mind. These chimps were our closest primate relatives—the most like our own species. Their violence was staggering and extremely frightening, like similar kinds of human behavior—rampages, unprovoked attacks, and the strong who single out the weak for violence—what we too often see on the nightly news. Perhaps we haven't evolved as much as we would like to believe.

In marked contrast to the male displays of power, it was soothing to observe the calm mother-infant relationships, like those of Madame Bee and her daughters. Most mother-infant pairs spend the majority of their time in relaxed and peaceful encounters, such as grooming, playing, and foraging. I had time to watch Madame Bee with her infant and see how her older daughter, Honey Bee, helped from time to time by grooming her mother. Honey Bee's real motive, though, seemed to be trying to get closer to her infant sister, as she constantly reached over to gently touch Little Bee. Honey Bee would even sniff the ground where her little sibling had sat, as if she were programmed to recognize even the scent of her infant sister. The intimate interactions and communications between these family members may have been one of the reasons Madame Bee could successfully raise her offspring in the wild even with her disabled arm.

By midafternoon, I decided to leave Rafael and Esilom and head back to Gombe on my own. Slated for duty at the camp medical clinic, I wanted to take a dip in the lake before my shift. I arrived at our make-shift one-room medical clinic in time to help during doctors' hours, then joined the other researchers for dinner at the dining hall, where I found out I was expected to give an account of my trip to the southern community. Jane, Richard, Nancy, Caroline, and Bill were curious to hear about the Bee family and how Madame Bee was doing with her infant. Sitting at a large wooden table, I described the brutal attack on Madame Bee as everyone around the table shook their heads.

"Why would Godi and Sniff attack Madame Bee?" one of the new students asked. "What was the point?"

Though no clear answer came from our group, I began to realize that all the adrenaline circulating in these powerful chimps—necessary to defend themselves and hunt for food—could also have destructive consequences when randomly directed. The attack had looked terribly fierce, but the chimps had used only a portion of their force against the Bee family. I knew they could have killed them outright.

After leaving Gombe, I learned about intentional attacks by the northern Kasekela males, who annihilated the southern Kahama

ABOVE: Rescued chimp Babu riding on the shoulders of Stanford student Lynne Johnson Davison on outing near the University. *Photo by John Crocker, 1973.* BELOW: View across the lush valleys of Gombe Stream National Park with Lake Tanganyika to the west. Home now to approximately one hundred chimpanzees. *Photo by Grant Heidrich, 1973.*

OPPOSITE: Winkle cradling infant Wilkie, who tries to reach out to touch seven-year-old Atlas. *Photo by John Crocker, 1973.* BELOW: Jane and me in the Gombe forest observing chimpanzees together during my students days. *Photo by Aadje Geertsema, 1973.*

OPPOSITE TOP: Prof hangs on to a branch as he plays close to Passion, who is seen in the background grooming her daughter Pom. *Photo by John Crocker, 1973.*
OPPOSITE BOTTOM: Freud with play face grabs Gremlin as Fifi (left) and Melissa rest. *Photo by John Crocker, 1973.* ABOVE: Alpha male Figan appearing on guard and powerful while eating prized bananas. *Photo by John Crocker, 1973.*

ABOVE: Gremlin, as an adult, nibbles on termites clinging to a hand-crafted twig that had been inserted and then withdrawn from a hole in a termite mound. *Photo copyright © The Jane Goodall Institute / by Jane Goodall.* OPPOSITE TOP: Miff with infant Michaelmas, in perfect position for nursing, staying warm and being protected, following a downpour. *Photo by John Crocker, 1974.* OPPOSITE BOTTOM: Adult male chimp patrolling from vantage point high above the valley. *Photo by Curt Busse, 1974.*

ABOVE: Our group of students and researchers. Back row from left: Curt Busse, Caroline Tutin, Grant Heidrich, Anthony Collins, Julie Johnson, Emilie Bergman Riss, and Jane. Front from left: Jim Moore, Lisa Nowell, and me. *Photo courtesy of John Crocker, 1974.* BELOW: Field assistant Esilom Mpongo and me talking about life before setting out to follow chimps. *Photo courtesy of John Crocker, 1973.*

males—the ones I observed attacking Madame Bee. This kind of "gang warfare" had not been seen during the first fifteen years of Jane's study. Were these vicious attacks on their own species also related to primal survival behaviors? There had been a few years of peaceful interaction between the two groups, but then the northern adult males began to patrol the border and make reconnaissance missions on the smaller southern group. Months later they started to single out individual southern males and attack them viciously. Eventually, they killed all the adult males and took over their territory. They did not, however, hurt the adolescent females, who eventually became part of their Kasekela community. Some of the adult females were killed and some likely joined other chimp communities.

These brutal attacks may have occurred as a result of changes in food and water supply or due to human destruction of the corridors through which chimps travel and widen their gene pool. They also may have occurred simply because of episodic behavior we don't yet understand. In a natural environment, aggressive behavior by single individuals—or even gang warfare—might be adaptive over time. It would allow the violent behavior to be emphasized in the gene pool. Barring some exceptions, the more-dominant males usually mate more frequently than the less-dominant males. These aggressive actions might be needed for the species to ensure that the "aggressive genes" are passed on to the next generation to help them survive the dangers in the wild.

At our table that evening, Jane was impressed that Madame Bee had survived the attack while still protecting her daughters. "Good old Madame Bee," Jane said. "I know she is strong to have survived polio."

That night I contemplated the range of chimp behavior I'd seen that day. From male violence to the determined female protection of the young, the similarities between our human family and our chimp relations were impossible to ignore.

CHAPTER SIX

TRUST AND SECURITY

By the time I reached the halfway point in my stay at Gombe, the other researchers and I had settled into a routine—if you can call studying a beachside baboon troop, as some of the researchers were doing, or scampering after chimpanzees routine. Along with our Tanzanian field assistants, we shared fieldwork, chores, conversations, and adventures. Spur-of-the-moment forest excursions to the waterfall or Rift Mountains also provided extraordinary highlights to the already amazing endeavor of tracking primates. On the one hand, I felt unique and privileged when writing home to my friends about hearing the thunderous sound of Cape buffalo and feeling the earth shaking below my feet. On the other hand, climbing a tree to safety to escape their approach, as I had been warned I should do, and nervously waiting for the giant beasts to move on made me fleetingly wish for the safe haven of a cozy college library cubicle.

In our potentially dangerous surroundings, the greatest sense of safety came from having my colleagues around me for support, bolstered by the thirteen-year history of safety that previous Gombe groups had enjoyed. My anxieties about the dangers were eased by the steady companionship of the field assistants during the day and by hearing the staff's drumming around holiday beach fires at night. Both gave me reassurance. And Jane's confidence in the world around her was like a sturdy anchor in a storm.

Still, who would think a young man could discover trust and security living in a small hut in Africa among wild animals, in unpredictable terrain, and exposed to tropical diseases and other hazards? Harboring vestiges of my dad's overall untrusting nature, I found myself gradually stripping off my paranoia and letting my surroundings provide comfort instead. This transition occurred over time at Gombe, the result of my widening forest experiences, and often with the help of close human companionship.

Because of everything from Jane's warm welcome when I arrived at camp to a researcher, Tony, reading poetry to me while I experimented with spending a night in an abandoned chimp's nest, my confidence to navigate in the wild forest grew steadily. Even today, I recall the comfort it brought me when Tony came by to check on me with his flashlight as I lay curled and cramped in the nest. The empathetic Scotsman in charge of the baboon study had known I might need reassurance, being high up in the swaying tree branches in the dark, and he began reading poems from the ground below. His Scottish accent echoed through the forest, and I was soothed enough to remain in the nest until morning. Tony had returned to the comfort of his human bed back at camp shortly after midnight, but his reassurance was long-lasting.

The field assistants certainly provided me with both emotional and physical security. Mostly men in their twenties from local villages, they were very kindhearted yet tough as could be. I watched them play soccer on the dirt with their bare feet. They were chosen for these prestigious positions at Gombe because of their keen observational skills and knowledge of the local flora and fauna.

When I first arrived, Esilom, a field assistant I came to know well, confidently pointed to a chimp fifty yards away and said, to my amazement, "That is Figan." I knew at that moment that my field notes and observations were going to be accurate thanks to Esilom, Hamisi, Rugema, Hilali, Yahaya, and others.

In the early years at Gombe, a female researcher had fallen to her death while in the forest alone, so it was required that students and graduate researchers be accompanied by a field assistant whenever they followed the chimps. I felt completely safe with all of them because of their expertise in understanding the animals and topography of the Gombe forest.

However, one night we were put to the test. Esilom and I headed back to camp as usual after our chimps got settled into nests several valleys away from the camp. After we had walked a while, we passed a tree that looked familiar.

"Hey, have we been here before?" I asked Esilom, pointing.

We stopped walking, and he looked at me and frowned. "Hm," he said. "Yes, maybe." I waited while he looked around a bit.

We resumed walking. It was growing dark, and the forest was very dense. The moon wasn't visible, and our one flashlight had only a faint glow of light remaining. Esilom shook it and banged it against his hand, but to no avail. I had a sinking feeling we might be spending the entire night in the forest without any walls around us for protection. My tired, hungry body was feeling a bit weak.

"Let's find the beach to get our bearings," Esilom said, but after we had walked a while in what we thought was the right direction, we seemed to be going in circles and getting nowhere.

I was worried about being lost, but at the same time I felt secure about Esilom's navigational abilities. *He'll get us back on track*, I told myself. "How are we doing, bwana?" I asked Esilom.

"*Sawe sawe*," he replied, meaning OK—just OK.

We were both more edgy than usual. When a nearby bush pig grunted loudly, we startled simultaneously, leaping side by side away from the sound. I felt awkwardly jumpy, tired, and hungry all at the same time.

The forest seemed to be getting even darker, and it was becoming more difficult to locate trails. An hour later, we ran into a creek. Esilom said, "Let's follow this! It will lead us to the beach."

"Yes, OK," I said, and we set off again. Scampering blindly down the valley next to the stream—and sometimes in the stream—I had no idea what I was stepping on. At least nothing was biting or stinging me between the straps of my plastic sandals. I was so relieved to know we could now find the lakeshore that nothing else seemed to matter. Our built-up adrenaline provided us with such energy that we essentially ran the entire distance to the beach. Going downhill also helped.

When we arrived at the beach, Esilom let out a laugh. Next thing I knew, we were both laughing so hard we had to sit down. We laughed for several minutes without saying a word, releasing more of the built-up tension. It was a huge relief to be on the beach with more visibility and the secure feeling that we would get back safely. Then we walked the rest of the way back to camp, planning what we would tell the group to save face. I favored a story about encountering several chimps treed by a circling leopard and having to stay to make sure they were safe.

After this intense adventure, some of my free-floating anxiety about jungle dangers seemed to dissolve; my restless fear of something happening to me dissipated. Perhaps I simply recognized for the first time that any time I was out in the forest, my safety would be my field assistant's paramount concern.

Of the ten field assistants, I was most in awe of Hamisi Matama. Hamisi gently guided me through the forest and gave me confidence in myself as a chimp observer. I admired his deep understanding of nature as well as his wisdom about life. Hamisi was always quiet and relaxed while following the chimps. He constantly detected sounds, smells, and movements in the forest as he pointed out specific birds, chimp behaviors, and a diversity of plant life. He was the first person I had encountered in my twenty-two years who was so extraordinarily sensitive to his natural surroundings.

Hamisi's calm in the face of challenging chimp encounters was made clear to me early on. Just two weeks after my arrival, the alpha male, Figan, and his brother Faben killed a colobus monkey, and I found myself in the midst of screaming chimps and baboons who had arrived at the scene hoping for scraps of the prized meat. Noticing the uncertainty on my face, Hamisi walked over to where I stood and simply nodded his head. Just having him next to me was all the reassurance I needed.

On our chimp follows, Hamisi knew the terrain as well as how to avoid hidden dangers. One day I was concentrating hard on the chimps as I ran along the trail. Suddenly I was jerked to a halt; Hamisi had grabbed me by the shirt to prevent me from going over a drop-off not far from the top of the waterfall. I looked down, my heart pounding with the exertion and the realization of how close I had come.

"We have to stop here and let them go," Hamisi said calmly, but with a concerned expression.

"Wow." I needed to be more aware of the topography on our fast-paced travels. "Thank you," I told him, as I heaved a deep sigh. I also needed to be sure to let him go first.

Hamisi's forest posture was something to emulate. He could be completely focused and home in on a single chimp while never losing sight of the larger environment. He had a kind of double vision, able to see both the forest and the trees simultaneously.

My fellow researchers added another level of trust and security. After a night of sleeping poorly, I awoke to find that my right leg appeared twice its normal size. I simply stared at it in horror for several minutes, but then finally hauled it out of bed. I struggled to get my shorts on, and then headed down to the dining hall for breakfast.

When I walked in, Emilie did a double take. "Whoa, what happened?"

I said, "I—I'm not sure. It looks infected." Several people came over to inspect the swollen, tender limb. Caroline—a chimp researcher—and Tony also looked worried.

"We're going to the hospital," Emilie announced.

"No, no, it'll be OK. I figured I'd just see how it did today," I told her. She looked unimpressed. I looked down at the leg. I didn't want to miss out on a day with Hamisi and the chimps. But it did look pretty awful.

"Come on, tough guy. We're going to the hospital," she said, and I reluctantly agreed.

It turned out I had a bacterial and fungal infection in the leg, caused by my plastic sandals rubbing my skin raw; it was a good thing that Emilie had made me go see the doctor in Kigoma. After my six-hour trip to the local hospital and back, I had to spend time recuperating in my hut. The worst of the treatment was a stiff brush the doctor insisted I use to firmly brush and cleanse the open wound on my foot where the infection started. It caused excruciating pain. But it worked! Bill, Caroline, and Tony brought me food and helped me care for the swollen leg by bandaging it and reminding me to take the handful of antibiotic and antifungal pills each day.

Though very shy about all the fussing over me, I gained a deeper understanding that this community was not going to let one of its members suffer alone. Nobody revered a self-sacrificing martyr when it came to injuries or illness. The community had to trust every member and be able to work as a group to care for each person. An untreated illness or injury could escalate to an emergency requiring heroic efforts or even impact the future of the program if its safety record was called into question.

One day while I was working in the clinic, a nine-year-old girl came in with her father. Julie examined the girl, and then called me over. She gestured to the expressionless child's firm, distended abdomen. I wondered if the girl had a parasitic disease with a possible blocked intestine. "John, would you take Rashida and her father to Kigoma? She needs to go to the hospital."

The town of Kigoma was three hours south of camp. The route was familiar since I'd already made several trips in the camp's small, wooden, gas-powered boat to buy food at Kigoma's markets, and our trip there together was uneventful.

I walked Rashida and her father to the local hospital, a one-story cinderblock building a few blocks up from the main part of town. The fact that it was 9:00 P.M., dark, and bedtime for most of the town did not keep people away, as a line of people waited in the holding area to be seen. Personnel in white uniforms looked very solemn and official, and we received short, crisp answers from the doctors and nurses, all of whom were quite busy with other patients. It took a while for her to be admitted, but once she was, I left them there, giving her father a reassuring pat on the arm. I headed the boat back to Gombe alone—in the dark.

The first part of the trip felt mystical. Ramadan was being celebrated in the villages, and all along the shore small fires flickered. Goosebumps rose on my arms as I listened to chants and songs echoing over the water. There was no moon. The sky was pure black, and the brilliant stars and fishing-boat lantern lights reflected in the lake. There seemed to be no distinction between sky and water. Even the temperature felt the same when I let my fingertips dip in. The blackness and lights and chants were hypnotic; I began to feel as if I were traveling through the Milky Way.

Eventually coming out of my reverie, I looked around and my heart leaped into my throat. I was lost. I could no longer see landmarks along the shore for navigating, and I imagined submerged rocks damaging the motor. In an instant, my serenity had transformed into panic. *Where is camp?* I wondered frantically. It was late, and everyone at Gombe was likely asleep. *Will it occur to anyone that I might be in danger?*

After what seemed like an eternity, I noticed that there were no more villagers' fires along the beach. That meant I must be getting closer to Gombe—or that I had already passed it.

Finally, at around midnight, I made out Jane's house just beyond the boat landing area. With a shout of relief, I turned the boat around and brought it ashore. I saw a figure approaching—it was Esilom. I felt like crying with happiness and relief when I recognized him. He came down to the water, helped me secure the boat, and made sure I was OK. Knowing I would be back late, he had waited up for me. A warm feeling of close friendship and brotherly support washed over me.

Though afterward I knew better than to try to navigate the lake alone at night, I had also learned to appreciate the "village" of trusted individuals around me. Esilom's conscientiousness and kindheartedness in watching out for me that night will always remain etched in my heart.

CHAPTER SEVEN

MY FRIEND AND MENTOR: HAMISI MATAMA

Working on such a remarkable project and living together in such a remarkable place created a strong sense of identification and community among the field assistants and researchers. We all felt extraordinarily lucky to be working with Jane Goodall, and I felt particularly lucky to be experiencing the world beyond my own. I made many friends, but of all the field assistants, it was Hamisi Matama who became my closest one.

Initially, Hamisi's field style puzzled me; I thought he was a bit detached as he gazed at the trees and animal life, showing little emotion. I felt engaged and exhilarated while watching the chimps—every nerve ending buzzed. Hamisi had a preternatural calm that seemed at odds with the emotional intensity of following and observing the chimps, but I soon realized that I was misreading him.

The turning point occurred on a trek one day a few weeks after my arrival. We were sitting quietly, side by side on the ground, and I thought he was daydreaming. Then, seemingly out of the blue, he looked at me and asked, "Bwana John, *unataka kuona nyoka*?" meaning, Do you want to see a snake?

We stood and walked about fifteen feet. Then he pointed to the ground to show me what I thought was a narrow log in the brush—until it started slowly moving. It was an enormous python! I startled and backed away as it crept on. Hamisi had been listening to movements in the grass and perhaps had caught a glimpse of the reptile as we were sitting together. I had misread his silence as boredom or disinterest, when in fact, he had been attuned to all the visual and other sensory cues in the surroundings. He was a superb tracker of forest animals. The huge smile he flashed at me as I scrambled away conveyed his humor and love for this arduous and rewarding work.

Hamisi's focus integrated everything he observed—including me. When we were hiking up the valley together, he seemed focused on the landscape, but he suddenly announced, "I think we should stop here and rest a while" when I had just started feeling like my legs wouldn't carry me any farther. This was not the only time when he picked up on my needs. In addition to noticing when I was growing tired, he seemed to know just when I needed help identifying a particular chimp off in the distance. He also understood the forest and how to avoid its perils—both seen and unseen.

There were, of course, cultural differences between us that could take me by surprise. Once, as we walked along the beach before dinner, the sun approaching the horizon, and men outfitting their small boats for night fishing, Hamisi said softly, "The men are setting out for fishing."

I was listening carefully and glancing down at the sand, so I was startled when Hamisi very gently clasped my hand in his. I had seen

Tanzanian men walking together holding hands, and knew it was part of their culture, but that didn't stop my face from flushing hotly. It seemed like forever until Hamisi let go of my hand, but it was probably less than three minutes. I was deeply self-conscious, but also flattered at the same time. Soon I grew comfortable with this tradition. I felt I was being accepted as part of the closely bonded "men's circle," just as the Tanzanian women had their own "women's circle." I enjoyed the natural ease of being emotionally connected to the field assistants, and I recognize now that as a result of my time at Gombe I am more open and connected with my friends, having brought some of the brotherhood style of interaction home with me.

Often while traipsing after the chimps with Hamisi as my guide, I thought about how fascinating it would be to visit Hamisi's home in Bubongo Village. This was a five-hour hike to the other side of the Rift Mountains. I hadn't heard of other students making such a journey to visit the home of a field assistant, but I was deeply curious. There is so much to experience in Africa—jungles, deserts, and mountains; rural villages and growing cities; diverse cultures, countries, and communities. I wanted to see and experience this one place: the rural village where my friend grew up, his particular Africa.

One afternoon, Hamisi invited me to come home with him. He evidently took the initiative after I'd dropped numerous hints such as, "What is your home like? Do you grow your own food? Do you have chickens or other animals? Do you live with your parents? What do young children do in your village? How far is the nearest school?"

The trip was one of the highlights of my time in Tanzania. I followed Hamisi along dusty trails, climbing high up into the Rift Mountains, gaping at the gorgeous vistas. After months in the dense forest, I was suddenly completely exposed to the sun as we strolled along the barren hillside. The warm breeze caressed my face. My imagination kicked in and I daydreamed of flying like a bird out toward the massive lake and looking back at the deep green valleys where the chimps lived. The open and stunning landscape had ignited my creative side as I pictured myself soaring like an eagle high above the Rift Mountains. The sky was deep blue against the soft-brown of the mountains, and not a single cloud was visible. Despite the arid climate on the mountain at 4,500-foot elevation, Hamisi picked a small white flower from a bush and had me smell the gardenia-like fragrance.

Then, once we began our descent, we passed a number of small corn and cassava farms that I otherwise wouldn't have seen during my time in Africa. Cassava was the main staple of the Tanzanian diet, a woody shrub with a starchy tuberous root that was pounded and cooked into a baked-potato-like consistency. The small plots of lush plants enclosed by wooden fences were close to a stream that winded down to the culti-vated valley below. We were definitely leaving chimp land and entering human civilization.

Along the path, we encountered an older man, whom I greeted by lowering my head and saying, "*Shikamoo mzee*," the way one acknowl-edges respect for an older person. However, probably because I was Caucasian, the man also lowered his head. I lowered mine farther; he lowered his further. As each of us then tried to lower his head more than the other, Hamisi began to laugh. His laughter grew louder as our heads nearly touched the ground.

Finally, we reached Hamisi's homestead, where five simple mud and thatch houses clustered on a verdant hillside. Beyond the houses I could see a large valley off in the distance, far below. Hamisi lived with his parents and younger siblings in one home, with another for his grand-parents, and one each for his older siblings and their families. A small garden adjoined them, and chickens ran about freely. As I walked across the rich, reddish-brown soil on a small pathway leading to the main house, I felt nervous not knowing how I should act or if my Swahili was good enough to carry on conversations.

That feeling vanished as soon as I met his mother. "Welcome to our village and home after your long trip," Hamisi's mother said graciously in Swahili. "Perhaps you would like some tea?" When I gratefully nodded, she excused herself to begin preparing the beverage as well as an afternoon meal.

After meeting more of Hamisi's relatives and taking a stroll around the garden, I sat on the veranda and visited with family members while the children played close by. Some of the younger kids peered out of a doorway and stared at me curiously. When tea was ready, Hamisi's mother invited me inside, where the earthy aromas reminded me of Indian spices and fresh clay pottery. She filled a teacup half with black, spicy tea, added some milk, and then ladled in enormous spoonfuls of sugar; the sugar must have taken up about a fourth of the cup. After stirring it all together, she carefully brought it over to me.

I tasted the syrupy drink and smiled, trying to ignore the shockingly sweet taste. I wanted to get it away from me, but I was stuck. *Drink it all,* I kept telling myself. *Just do it.* The task took a while to complete—and I felt a bit nauseated afterward—but I wanted to be a gracious guest.

Soon it was time for our midafternoon dinner—a wonderful meal of cassava, beans, fish sauce, and fresh greens, all prepared over a small fire and without running water. Having worked all day, the fourteen-member extended family was ready to relax, talk, laugh, and simply be together.

"Hamisi told us you are studying to become a doctor?" his mother asked.

"*Ndio,*" I replied, nodding. "It will take four more years of schooling and three years of specialty training before I can practice medicine."

"Mmm—that is too long. You could work here as a healer even now if you are good at healing. Hamisi has a good understanding of plants that are used for curing illnesses."

I sensed that most people in the villages had better access to local healers and often preferred them to the government-run clinic in

Kigoma. I didn't know if the children had received immunizations to protect against polio and other illnesses, but everyone looked healthy and fit.

After I'd finished my plate, Hamisi's mother wanted to pile it with more food. Thankfully, I had learned a useful Swahili phrase for this situation. With a smile and slight bow of my head, I said with confidence, *"Nimeshiba,"* which means, "I am very satisfied and cannot eat any more." It's a lovely one-word way to politely express satiety. The meaning is imbued with appreciation for the meal. I thought gratefully of my Swahili teacher, who had taught me this crucial phrase, the only polite way for a guest to refuse more.

The afternoon was filled with easygoing, joyous interactions among Hamisi's extended family, mostly talking and walking around the garden, dodging the chickens, and watching the kids play. It made me wish I had grown up with a similar style of togetherness. My own upbringing had not created these warm familial bonds, though I felt a close connection with my mother, who always anticipated our needs and was available for reassurance. Though she prepared nice, wholesome meals as a stay-at-home mom, what I remember most is her opening cans of Franco-American spaghetti, my all-time favorite!

But the relaxed pace and inclusiveness in Hamisi's home contrasted with the more strained family life I had known around dinnertime, our main shared meal. I grew up with a father who was painstakingly working his way up the corporate ladder. He frequently came home after work late and in a cranky mood. The four of us kids managed to find humor and entertainment among ourselves as we tried to evade his venting of pent-up stress.

After my Bubongo visit, I wondered if, when I went home, I could persuade my parents to adopt a predinner ritual of relaxing and just being together—but in the winter months in Minnesota, it certainly wouldn't take place on the front porch.

As I walked back to Gombe with Hamisi that evening, I realized how spiritually close I felt to his community. People living in Bubongo seemed to have a strong connection to nature, and they recognized how much I valued this. Without electricity, people spent much of the time outdoors. Traveling by foot was a welcome change from the energy-draining traffic back home. I was also attracted to the slow pace at which people interacted and prepared their meals. Food, family, and time were all woven together. It would have been inconceivable for Hamisi or his family to rush through a meal or put limits on their companionship.

Perhaps I was searching for a model of home life that included more mindfulness and thankfulness. Even if I was seeing the best of family life on display in Bubongo, that didn't really matter. I was creating a romantic and charming view of how my future family would be. I found myself wanting to return to the village and spend the night. I pictured myself sleeping on a thin mat a few inches above the dirt floor as I listened to children breathing nearby, the night wind, and the morning roosters.

Over the next months that we worked together, Hamisi and I became even better friends, but despite our closeness, I had to think twice about my urge to invite him to join me on my three-week break to climb Mount Kilimanjaro. There were delicate social rules to consider. At Gombe, it wasn't customary to travel with a field assistant, except into town for a food run. Perhaps this was because it would single someone out and create jealousies, or perhaps because of the expense or communication difficulties. In the end, I didn't care about these issues—except for the cost. I would be paying for both of us and couldn't afford to visit big-game parks or stay in even moderately priced hotels. It would be third class all the way. But with my friend Hamisi as a traveling companion, language and cultural barriers wouldn't be a worry.

"Would you like to climb Kilimanjaro with me?" I asked Hamisi one afternoon while we sat together in the upper camp.

He looked shocked and then gave me a huge smile as he realized I really meant it. "*Asante sana,* bwana," he said, thanking me for the invitation.

For our "relaxing" vacation, Hamisi and I traveled third-class by train for two and a half days to Dar es Salaam, as the first leg of our trip to Moshi, where we would climb Kilimanjaro together. We shared our standing-room-only train car leaving Kigoma with chickens, goats, and a lot of other people. I was packed in so tightly I couldn't possibly fall over, and the fragrance of the animals in our compartment added another earthy edge to the journey. Luckily, Hamisi brought along some fruit and bread, and I had stuffed some cooked chicken into my backpack, which we devoured. We had wisely planned to sleep in a small hotel halfway to Dar, and getting off the train for a while was a welcome relief.

Tony, the Scotsman in charge of the baboon study, had had the original idea for the three of us to climb the mountain. He met us at its base, and we joined a motley group of twelve people from all over the world along with a very experienced guide.

Starting in a tropical valley, we passed through rapidly changing vegetation up to ice and snow, which covered the ground for the final 4,000 feet up to the summit at 19,300 feet. Hamisi slowed down when we began trekking near snow drifts, and I suddenly realized that snow was completely foreign to him. I turned around to see him grinning and holding the snow and crumbling it in his hands. I bent over, scooped up some cold handfuls of snow, made a snowball, and threw it at him. It might not have been the best timing, because he was still exploring the finer details of the snow. Instead of laughing, he looked at me as if he didn't understand why I'd suddenly pelted him.

"It's what we do when we play with snow," I explained to him. "We call it a snowball fight!"

Hamisi raised his eyebrows in uncertainty. Snow was such a marvel that using it in play fighting was not the first thing he would have imagined doing.

We got colder and colder as we climbed; our light packing job and borrowed gear were woefully inadequate for the crisp, biting air at higher elevations. The last night of our four-day ascent, both of us were so cold that we barely slept. We were also experiencing altitude sickness, he more than I. He had grown increasingly quiet, holding his stomach and wanting to lie down. He said he had a headache and felt weak. For the final ascent, which began at 1:00 A.M., Hamisi stayed behind.

"You should go on without me," he said softly. "I am too cold and don't feel well." I could tell he was experiencing the effects of the elevation and plunging temperatures, but he'd made it through the most enjoyable portion of the climb, the first fifteen thousand feet, for which we'd had blue skies and reasonable temperatures. I regretted having to leave him at camp, but knew that it might be risky for him to continue.

Tony and I finally reached the summit at 8:00 A.M. My stomach was rebelling, my entire body was cold, and I was short of breath. Clouds enveloped us, concealing the view. I glanced grimly at Tony and said, "Why are we doing this? Was it your idea or mine?"

The scenery had been so interesting lower down, and we'd all felt comfortable there. Here at the top, I did feel a sense of accomplishment, but I also felt guilty that Hamisi was stuck, shivering, back at the last camp. Maybe if the sun had been shining on the snow and I could have seen for miles across the Serengeti, I would have been able to muster more excitement, but nauseated from altitude sickness and seeing only clouds from the peak, I made a discovery about myself: my rewards in life come more from everyday accomplishments than splashy feats such as conquering mountains. I felt prouder about being Hamisi's friend than I did about summiting Kilimanjaro.

I also paused to consider the driving force within humanity to "reach the top." The same ingrained trait of dominance that kept Figan challenging his way to the top of the chimp hierarchy was probably at work in each of us on the mountain. To reach the top had seemed important while we were climbing. When George Mallory was asked why he had wanted to climb Everest, he answered, "Because it's there." But if I were to return to Kilimanjaro today, I'd be content to hike to the fourth hut on the mountain and forego the final four thousand feet. I would move more slowly, taking time to notice the changes in the landscape as we ascended and paying more attention to my companions and the mountain itself.

I stayed at the summit, peering off into the clouds for five or ten minutes, and then headed back down, happily reuniting with Hamisi, who was feeling a bit better from resting and starting to acclimatize. The next day, we ran down the mountain, since there was no need to slowly adapt to decreasing altitude—at least at our age. Our rented lightweight boots were wet and uncomfortable, but we weren't too concerned. Though we had not been well prepared for the cold or the altitude, we were young, resilient from our long treks in the forest, and still learning about nature and our own survival skills.

After the arduous climb, Hamisi seemed anxious to get home. At the very least, he didn't express enthusiasm about seeing more towns and was still feeling weak after the climb. He had been very cautious on the entire trip, from what food he sampled to talking to people along the way. His one big interest, I noticed, was purchasing a few *kitangas*, African printed fabric that women wear in wrap-around fashion, to bring back to his village as gifts.

Looking back, it makes sense that Hamisi was generally hesitant and reserved. Even in his own element he was a quiet and contemplative person; he approached everything with care. He also had never traveled to a major city, especially with a white guy who was still learning Swahili. At the time, I wondered why he didn't seem more excited about the variety of life and merchandise in Dar es Salaam, where we had spent time before the climb. It may have been because it was overwhelming his sensitive nature. Compared to Bubongo Village, where there were no cars or electricity, a major city such as Dar es Salaam with traffic, fast-moving people on the sidewalks and plenty of noise from trucks and factories might have seemed like a war zone to Hamisi. But it could also have been, in part, that he was enthused but just wasn't meeting my expectations of how excitement should be expressed. I came from a culture where excitement and accomplishment were often expressed by fist pumping and loud shouting. I was more familiar with the kinds of energetic displays I saw from Frodo and Figan as they worked their way to the top of the male hierarchy. Hamisi may have been experiencing excitement and joy while we were there but not needing to display it outwardly.

Sharing that adventure with Hamisi meant a lot to me, and I found out afterward that it also meant a great deal to him. Back at Gombe, several days later, I was compiling some data in the main lodge. He entered with a basket, looking full of anticipation. He presented it to me; it was fifty hard-boiled eggs. "Oh, Hamisi!" I exclaimed, astonished and touched. This was a huge gift and expense for him, and I was deeply appreciative and humbled. It took me a month to eat all those eggs—even sharing some of them. I thought about our trip with each bite.

Perhaps tellingly, the trip to Hamisi's village remained more imprinted in my memory than climbing Mount Kilimanjaro. Topping a mountain may have given me bragging rights, but being so warmly embraced by Hamisi and his family was the real and lasting pleasure. Hamisi had been very proud to introduce me to his parents and other relatives. I appreciated experiencing unconditional acceptance and warmth in a faraway place during a time of searching for meaning in my life. That being said, climbing Mount Kilimanjaro together was an important milestone in my growing friendship with Hamisi. Shared adventure and shared adversity are important bonding rituals. These do not leave us unchanged, and we change together through them.

There were many aspects of Hamisi's life that I didn't learn about until my return thirty-six years later. I looked to him as a mentor because of his expertise in tracking animals, his knowledge of plants, his understanding of the chimps, and his extraordinary abilities in the forest. I later found out that, even though he appeared to be a few years older than me, he was actually five years younger. He was only seventeen when we worked together. Seventeen. I also did not know that his traditional name was Mlongwe. I'd never heard him called that, so I assumed he had been named Hamisi at birth. Mulongwe is also the name of a town on Lake Tanganyika, not too far northwest of Gombe in the Congo. Perhaps he was born there. The name Hamisi, which is a Swahili boy's name meaning "born on Thursday," was likely a nickname, though it's a very popular name itself.

Growing up in a remote village without running water or electricity, Hamisi likely learned survival skills and independence at an early age. His humble manner, which other field assistants possessed but not to such a degree, may have arisen from living in the shadow of his older brother, Hilali, who was also a field assistant at Gombe—but Hamisi's skills, abilities, and keen intelligence were all his own. When I found out his real age, I could only marvel at his maturity and think back to myself at seventeen. I could never have offered someone advice or guided them through a forest—or even a neighborhood, in my case. I had not been as composed, and easily blushed with the least bit of embarrassment, yet Hamisi and I shared common elements in our temperaments, such as sensitivity to our surroundings and wanting to be inclusive of the people around us.

During our time together, Hamisi and I became long-term friends, communicating nonverbally with knowing smiles, gestures, and glances, and transcending language difficulties. We had an innate understanding of each other and were able to work together as equals. We had established a strong sense of trust and familiarity. Though we didn't groom each other as the chimps did, our bond was nevertheless profound and comforting.

CHAPTER EIGHT

KNOWING JANE

As I immersed myself in the life of the Gombe chimps, I gained a deeper understanding of Jane and her work. At times I even questioned my plans to go to medical school because of my growing passion for spreading Jane's message about the importance of these wild primates.

Often, as I listened to Jane tell her stories at Gombe, I remembered our early experiences together. I had met with her for the first significant amount of time in the spring of 1972, soon after being accepted for the Gombe program. I was with a small group of students and faculty from the Stanford Human Biology program, gathered in the living room of a house in Palo Alto to discuss Jane's research. Still awestruck by her presence, I wandered into the kitchen after lunch and before long found myself doing dishes with Jane and her mother, Vanne. Standing beside the mother-daughter team, witnessing their comfortable working rhythm, I easily pictured them setting up camp together at Gombe in 1960. This was my introduction to the Goodall teamwork that prepared me for my work in Africa.

Vanne had been visiting from England to help Jane with her busy schedule and care for Jane's then-six-year-old son, Grub. I felt an immediate bond with Vanne, similar to the connection I later developed with Jane. She was very approachable, with a warm, welcoming, relaxed manner. Vanne usually remained in the background, attending to organizational details at these get-togethers, but she was always receptive to conversation, which I enjoyed immensely.

"How did you like living in England?" she asked, knowing I had spent time just southwest of London.

"I enjoyed the pubs and the people a lot." I instantly regretted putting pubs first—not wanting her to think beer was the main thing on my mind. To my relief, she gave me a knowing smile.

Before I left for Africa, I spent more time with Jane at department picnics and meetings. Getting to know her a bit through these events, I wondered if both Jane and Vanne had developed more nonverbal communication skills than the average person as a result of spending time with the Gombe chimps. I enjoyed this feature of their communication and tried to integrate some of it into my behavior. In a moment of illumination, I realized that my mother used a lot of nonverbal communication too.

A subtle smile by Jane or Vanne with direct eye contact meant "I agree," or "Good work"; Jane's hand on my knee while she was leading a discussion meant, "I like your sitting next to me even though I can't talk to you right now"; and a twinkle in Jane's eye meant, "I am happy"—better than mere words could express.

One of the most remarkable things about Jane was her ability to listen. This sounds far simpler than it was. As an accomplished listener, Jane was able to take action by doing nothing other than watching and listening. The idea that the act of seemingly doing nothing could actually be doing a great deal was revelatory. Listening and watching were themselves forms of communication. My own patterns of communication seemed overly verbose to me, so I tried to emulate Jane's more nonverbal style and keen observational skills, even after I'd left Gombe.

Later in life, this would influence my practice of medicine. As a doctor, I learned I could elicit better information from patients more efficiently if I talked less and listened and observed more. Channeling Jane, I would wait patiently and practice being receptive and observant. Rather than look at my computer screen and begin with, "I see you're here for knee pain, is that right?" I would make eye contact and greet patients with, "How are you doing?"

Jane and her mother impressed me with the apparent ease with which they navigated through life. I had read all I could about Jane before leaving for Gombe. I knew about her life growing up in Bournemouth, on the southern coast of England, and about her childhood love of animals. I knew she had shared her research and stories through films, books, and riveting lectures, inspiring people across the globe. She continues to write, lecture, and work tirelessly to encourage humanity to care for all living things and to protect their—all of our—habitats.

In the first few years of her study, Jane withstood criticism from the scientific community for describing the chimpanzees as though they might have emotions, and for naming them instead of numbering them. David Greybeard should have been chimpanzee #1 and Flo, chimpanzee #3. But Jane's approach helped people around the world to vividly picture the community of chimps at Gombe. Using the same letter for names within the same family, such as the F family, with Figan, Fifi, and Flo, helped people remember family lineage. It allowed people to connect with her work in a way that was enlightening. Chimps were not just cute and funny humanlike creatures in a zoo. The study of chimps wasn't just a specialized science that had nothing to do with humans. Instead,

Jane educated us about our connection to our primate cousins and the environment. Watching the Gombe chimps interact enabled us to see our own world through a new lens, and helped us recognize some of our own strengths and weaknesses. People learned to empathize with the chimps, remembering their names, their individuality, and their lives. For example, many years after the release of one of Jane's Gombe films, Jane was approached by an elderly woman in a small town in China who asked, "Please tell me more about Flo, whom I admire so much!"

During her earlier years, Jane witnessed previously unknown chimpanzee behaviors that were both exciting and disturbing. She watched David Greybeard modify a branch to fish for termites in their mound, an observation that changed the prevailing notion that humans were the only primates who made and used tools. Jane also watched young chimps learn to build complex sleeping platforms in the trees. And in perhaps one of the most important discoveries, she observed signs of severe depression over the loss of a family member. This added much to our understanding of the complex emotional and intellectual capabilities of our primate cousins. We were able to fully recognize these humanlike reactions in chimps, deepening our knowledge of not just the chimps but ourselves as well.

As I mentioned earlier, in 1974, Jane was stunned to witness warfare between the male chimpanzees from Kasekela Valley and the southern community of chimpanzees dwelling a few miles south, in Kahama Valley. In discussing these attacks during one of her talks—attacks like the one I witnessed against Madame Bee and her daughters—Jane commented, "I used to think that chimpanzees were nicer than humans, but now I don't think so."

Even after fifty years of observations, the researchers at Gombe are still witnessing behaviors and events never observed before, including the chimp Gremlin's feat of raising twins, Golden and Glitter, to adulthood in the wild. A few other pairs of twins were born at Gombe over the decades, but only one of the twins in each set survived past the first few years. This was because of the enormous energy and favorable circumstances required for success, such as having an older offspring to help. Gremlin was the first to accomplish this mother-infant success story. It's hard to imagine her nursing twins as she searched for food, built a larger nest for three chimps to sleep in each evening, and carried them through the forest for the first two years of their lives. But do it she did.

At Gombe, Jane was most relaxed and at home. It was at Gombe that I saw her laugh, joke, gaze, wonder, and seem completely comfortable—more so than during her teaching in California. At Gombe, Jane could maximize her many talents in bringing people together in a cohesive fashion. I loved the way she included everyone in our group in her plans and activities. Her manner of speaking was so thoughtful that anyone of any background could connect and be enthralled. And I especially valued the times when just the two of us talked. I considered these times my "audience with the pope." I wrote in my journal:

> When she speaks, people listen. Jane articulates her thoughts so well that she instantly creates images in the mind of the listener, with both the details and the big picture clearly understood. Even when working side by side with her, I can't figure out how she does it. I even tried adding an English accent when describing different chimp behaviors to my colleagues, but concluded that her speaking ability is more an innate talent than a learned one.

I felt a strong connection to Jane even when I wasn't around her. I thought of her while I was following the chimps, picturing the long days she spent observing them from a distance in her early research. I imagined her patiently waiting for the chimps to finally let her observe them at close range. Most people would have given up after a few months of lonely waiting in that perilous setting, but Jane's patience was rather like that of the mother chimps themselves. Patience, along with her determination and open, inquisitive mind, allowed Jane to recognize and interpret what others might merely have recorded.

One afternoon at the upper camp, midway through my stay, I had the opportunity to talk with Jane about the mother-daughter pairs I was studying. It felt like a chance to connect as both a friend and a researcher. As I stood next to Jane in the shade, discussing the chimps, she shared information about the behaviors of specific chimpanzee pairs, and, as is her style, she listened to my thoughts and ideas with total focus.

Jane described Passion's nine-year-old daughter, Pom, and Pom's reaction to the birth of her younger brother two years earlier. "When Prof was born," Jane explained, "Pom exhibited signs of depression and showed little interest in her young brother, who was stealing Passion's attention." I nodded, squinting and thinking, and she continued. "I think Pom is now finally getting more involved in the care of Prof—and even starting to protect him."

"I wonder if Passion's somewhat aloof mothering contributed to Pom being more needy when Prof entered the family," I cautiously offered.

Jane was always careful not to draw conclusions with just one or two anecdotal examples. "Why don't you see how Pom continues to do with her little brother while you're here at Gombe, and maybe we can understand it better over time," she reasoned.

I felt a surge of energy from Jane's open-mindedness and sense that there was more to learn. I vowed to monitor this family closely. As we talked, I realized this was probably the first time I had felt fully myself around Jane, without feeling inhibited by her fame. Looking back, I think she might have created this time in her wildly busy schedule to get to know me a bit better, which still means the world to me. Perhaps being the patient, listening mother that she was enabled her to recognize just what her example meant to me, and because of the openness she modeled, I was willing to relax and truly be myself, knowing she accepted me as I was.

While everyone at the camp appreciated the opportunity of working with Jane, a few of the Stanford students and other researchers challenged her on social and political issues regarding power and colonialism. A big controversy while I was there was whether the Tanzanian workers at the camp should provide us drinking water by carrying our hefty water buckets from the lake up to our huts once a week. Some students struggling with the Vietnam War and other political issues of the era thought we should carry our own water. I didn't have a strong opinion either way. I admit that this was typical of me at that age; I wanted to avoid conflict in Gombe, just as I did in my family while growing up.

"I feel like a colonialist when the Tanzanians carry our water buckets up to our huts," Bill stated in a secure and clear voice. "I think we are all strong and capable enough to do this ourselves."

"They've done this graciously for the past fourteen years," Jane said.

After spirited discussions, we did try carrying our own water for a few weeks; however, the Tanzanians felt bad when they saw some of the students, especially female students, hauling water up the path from the lake. The Tanzanians had taken pride in being strong and doing this work. In their own culture, the women carried the water to their homes in very large containers that they usually balanced on their heads, but at Gombe, the Tanzanians may well have been respecting what they had observed as a traditional Western view of female roles and duties—which didn't include carrying heavy water containers. Ironically, the discomfort some students felt in being "waited on" by the Tanzanian staff was replaced by the discomfort the Tanzanians felt when we took over these duties.

In the end, I think Jane understood the local Tanzanians, their customs, and their beliefs better than most people born outside the country. When the researchers decided to haul their own water, the Tanzanian camp staff who had been responsible for the task felt a loss of purpose and were also concerned about a loss of income. Jane understood how deeply a sense of purpose is tied to a sense of identity and belonging, which helped us resolve our dilemma—we eventually returned to having the Tanzanians carry water to our huts.

I was privileged to see many other facets of Jane during my time at Gombe. I discovered, for example, that she had a keen sense of humor. One afternoon, a group of us found ourselves trying to eat cucumbers with our toes after Jane instigated a contest. She could reach her mouth with the cucumber slice held between her big and second toes. She was not the only one at camp to possess this skill—but I was at least a foot away from the goal.

During our trip to Mount Kilimanjaro, Jane invited Hamisi and me to stay with her and her husband-to-be, Derek Bryceson, at her home in Dar es Salaam. Derek was the director of Tanzania's national parks. Jane was always gracious, but on this occasion she seemed to accomplish the impossible as a host with an unusually full house. Hamisi was not used to socializing with this group, and I was also a bit nervous. I had no idea how Derek would feel with all of us in the house during his budding relationship with Jane. In addition, Jane's son, Grub, was there, along with Tony, the head of the baboon study and Jane's right-hand man at Gombe. Jane helped all her guests feel comfortable by quietly making suggestions such as, "John and Hamisi might want to go with you, Derek, to snorkel off the boat. I think they would appreciate the beautiful tropical fish." We appreciated the way she directed us to activities she thought we would enjoy the most. After this, I always felt welcome in the Goodall household, whether at Stanford, Dar es Salaam, or Gombe.

One day, after I had become more familiar with everyone, I found myself driving the park boat back from Kigoma to Gombe with Jane and Derek. Jane and her first husband, Hugo, had parted amicably, and her fondness for Derek was growing. Seated next to each other, they held hands and sipped plum wine. They spoke to each other a little, but seemed more content just to smile at each other with dreamy looks in their eyes. It was fun being close to their romance, and it was heartwarming to see Jane so happy. I could understand why she was attracted to Derek. He was a true gentleman, with a warm smile and a gentle spirit. With his rosy English cheeks and bright blue eyes, he also had an inviting manner. Most impressively, he had character, along with discipline and determination.

Born in China and educated in England, Derek enlisted in the Royal Air Force in 1939 at the age of sixteen, and immediately became a combat fighter pilot. Three years later, his plane was shot down and he suffered a shattered pelvis and legs. His doctors told him he would never walk again, but, to everyone's surprise, with fortitude, endurance, and the use of crutches, he was eventually able to walk slowly.

Derek's romantic spirit was in full display one day when he flew his small plane from where he lived in Dar es Salaam over Jane's house at Gombe, and dropped a small package into Lake Tanganyika. Not knowing what was inside, Jane swam out and retrieved the rose he had wrapped for its free fall into the lake. Derek set a high standard of romance.

I was always amazed—during my student days at Gombe and in the years since—at how Jane made herself available to us and at her ability to stay connected with people, despite the demands on her. When Jane left Gombe for a while for one of her teaching sessions at Stanford, she occasionally sent handwritten letters to me at camp, describing both the excitement and challenges of her busy schedule. One of the most meaningful letters began:

> I'm in the midst of a whirl of activity, which is fast gathering momentum. Gombe group meetings, office hours, research assistant and intern meetings, categorizing slides and stills,

sorting notes, talking with students, visiting L.A. to look at baboon films, lecturing, and so on! Not to mention visiting Babu, which is another reason I had to write to you.

Jane knew of my concern for my old pal Babu after he had left his human family and entered a large compound with other chimps. Despite her many commitments in California, she was kind enough to let me know in writing how he was doing. She continued:

Babu is doing so well. I took a camera especially to get some pictures for you. Unfortunately, the group [of chimps] had many escapes the day before and so had been shut in all day for the repairs. Thursday, since it was a moonlit night, they all stayed outside half the night and Larry [the observer] watched them. Anyway, to see Babu climbing with Bashful, playing with Topsy, wandering off on his own, quite independent—it's really super.

It's hard to convey how influential Jane was to me during my time at Gombe. I was just one of thousands of students who found their lives shaped by Jane as a scientist and a human being. She continues to influence the way I view our world as a complex, interwoven landscape of species, cultures, and the codependency of all living things. Through Jane I learned to recognize the fragility of our planet and to understand how peril to one species is peril to all. Today, so much of the natural world seems to be in peril—and what we humans destroy might not ever return. But perhaps one of the most lasting life lessons I learned from my time at Gombe and from Jane herself was the title of her memoir, *Reason for Hope.* As a doctor I know that hope can be a powerful incentive to a patient—and a powerful part of how I practice medicine. Educators like Jane give me hope for humanity and for the world. Whatever our individual circumstances, we all need hope to survive and to meet life's challenges.

CHAPTER NINE

LEAVE-TAKING

After six months at Gombe, I had become much more relaxed and secure with my fieldwork and at ease in my friendships with colleagues. Most importantly, I had acquired a deeper understanding of chimpanzee behavior. Beyond the knowledge I had gained observing mother-infant pairs, I was especially intrigued by the chimps' capacity to express emotions. In the reassuring embraces and kisses of adults, the depressed behavior of young chimps in times of loss, and the joy of mothers and infants at play, I recognized feelings very much like mine. Though scientific doctrine taught all of us not to anthropomorphize the chimps, after observing them for many months, I concluded it was just as misleading to assume our two species didn't share similar capacities to feel and express emotions.

Later, after leaving Gombe, I would learn more about studies on chimp emotions. I read about the work of Dr. Roger Fouts, who taught the chimpanzee Washoe to use American Sign Language and revealed the depth of complex emotions that chimps experience and can communicate to humans and to each other. Fouts described a young woman who was learning sign language with Washoe but who needed to leave suddenly because she was having a miscarriage. When she eventually returned to see Washoe, the chimp ignored her. After the student signed for Washoe the reason that she had left, the chimp placed a finger below her eye signaling crying and sadness for the loss. The student was astounded by Washoe's empathy and ability to communicate emotions about this complex circumstance.

My own emotions kicked into play as the holidays approached and thoughts of friends and family naturally came to mind. Still, the lack of a familiar winter season prevented me from feeling melancholy or experiencing a strong urge to celebrate. By mid-December, however, some of the students had decided we needed some kind of Christmas-Hanukkah event, and so eight of us enjoyed a peaceful Christmas Eve with handmade gifts, walks along the beach, swimming, and a huge feast, including wine, duck, stuffing, mushrooms, cheese, and chocolate—luxuries we hadn't tasted in months—bought by one of the researchers while on break in Nairobi.

Though I enjoyed the festivities with my tight-knit Gombe family, they were eclipsed early Christmas Day when I opened my hut door to see Fifi and Freud playing quietly just ten feet away. Chimps had never come this close to my hut. Freud made soft grunting sounds as Fifi gently tickled and wrestled with him. I could almost reach out and touch them, but of course I didn't. I moved away to watch them from a distance, thinking about the extraordinary setting, of my entering the chimps' world and not even causing a stir, thanks to Jane. As birds chirped all around, I was overcome by a feeling of connectedness to the entire community—the people, the forest, the lake, and these remarkable chimps. At that moment, any holiday homesickness eased. The warm sun relaxed me, and my two favorite chimps stole my heart.

As New Year 1974 dawned, I thought my time at Gombe was coming to an end, but just as I was preparing to return to the States, I was asked to stay on to fill in for a student who had to leave early. Without hesitation, I agreed to remain until his replacement arrived several weeks later. In fact, I was overjoyed to extend my stay.

This extra time enabled me to join Tony and three other researchers on an afternoon hike to the barren Rift Mountains, which form the eastern border of Gombe. With sleeping bags in tow, we enjoyed a memorable campout, sleeping on the bare earth high above Gombe with bright stars overhead. What stayed with me most of the night was my sense of feeling the divide. To the east, we could hear the voices of villagers and see their fires down in the valleys. To the west, we could hear chimp calls and baboon grunts. Then, after nightfall, we heard only the wind. I was resting between two worlds: the world of modern humans and the world of ancient primates. One world was changing rapidly, and one had likely changed very little over millions of years. I would soon leave the world of my ancient primate cousins and reenter the world I knew best. I would need to adjust.

In the morning, we all hustled back to camp to start following the chimps and baboons we were studying and feel the connections with the forest and the wildlife around us. It would not be long before all of these connections would become just memories, and I carried the weight of that knowledge with me as I went.

By February, I had become one of the most experienced student researchers at the camp, but I was also about to leave. The time seemed right for my departure. Jane too would soon leave to travel in the States. I would start medical school in Cleveland in six months and needed to begin preparations for that transition. I knew it was time.

During my last week at Gombe, I was allowed to follow the chimps without completing the usual check sheets; I could just relax and feel the pleasure of being a guest in their community. One afternoon, I was fortunate enough to find Fifi in a neighboring valley to do my final follow with her and Freud. They and several other chimps were feeding on ripe mbula fruit.

Very quietly, Fifi and Freud descended from their treetop perch and headed down a streambed into the lower part of the valley. The late-afternoon sun was soft and golden as Juma and I followed them to a location with sturdy trees whose flexible branches were ideal for nest building. Calls echoed across the valley from other chimps searching out nest sites.

While Fifi began to build a nest for herself and Freud, I watched the sun setting across the lake behind the hills of the Congo. As the sky turned a brilliant reddish purple, Fifi slowly rolled into the nest and sprawled on her back. Freud joyfully nestled in next to her warm abdomen and suckled at her breast.

I remembered the first time I had seen Fifi during my early days at Gombe. What had struck me most was the rich black color of her shiny hair, her strong, healthy appearance, and her agility. Observing her dark form against the deep green of the forest, I could appreciate the power and vigor of the fifteen-year-old primate mother. The pair nestled in for the night, creating a picture of warmth and security unmatched by anything I could imagine until I had my own children.

Fifi and Freud would rest in their leafy bed suspended high in the trees until sunrise. Fifi's older brother Figan, and Melissa and Gremlin were within voice range in nearby trees. Safe from roaming animals and snakes below, they appeared well protected and comfortable. All was well with the chimps. But a sudden sadness hit me in the gut, a glimpse of the home-sickness for Gombe that I knew would begin as soon as I left this magical forest. In that moment I knew that someday I would have to come back.

As I made that last long hike away from Fifi and her family, I knew also that I would be entering a new phase in my life—one that would require me to be buried in books, indoors, and away from my friends of the last eight months—friends who had changed the way I viewed the world. I would miss everyone and everything about Gombe.

Though it was customary for the Tanzanian staff at Gombe to hold a farewell beach fire for people who were leaving, I assumed they thought my prolonged stay had served as the farewell party. I was mistaken. As I headed into the changing room near the beach after a lake swim, I ran into one of the field assistants.

"Bwana John, we are sorry you are leaving tomorrow. You must return someday. See you at the big fire tonight."

Sure enough, they built a huge fire on the beach and eighteen of us danced to the conga drum until we couldn't dance anymore. I remember seeing smiling faces pouring with sweat, lit up by the fire, moments before we jumped into the shimmering lake for a long rinse-off. My sadness about leaving was somewhat muted by the camaraderie of our celebration.

Finally I was ready to retire to my hut for a last night in the forest. When I reached my hut, that strange, lonely feeling returned, but I was so tired it was all I could do to blow out the candle before drifting into a deep sleep.

My departure from camp the next day was far from routine. Derek escorted me all the way to Dar es Salaam. I felt about as cool and impor-tant as anyone my age could feel traveling with Tanzania's national parks director. I realized also that it was only partly coincidental that Derek and I were both leaving Gombe at the same time. Jane had likely orchestrated the whole thing to help make my trip easier.

"Are you ready to go?" Derek asked once I was standing by the boat. He could see that I was, but was really asking the bigger question: *Are you emotionally ready to leave Gombe?* Derek was more tuned in than I was; as was typical for me, my sadness didn't fully register at the moment of departure. Instead, it hit full force after the boat was two miles away from Gombe, when I felt the deep ache of true sorrow.

We made our way by boat, small plane, and truck to Arusha, where Derek drove me to visit a vast park called Tarangire. "Our department just established this new game park," Derek revealed, with an obvious sense of accomplishment. "I expect it will become a major tourist attraction in the next few years."

I felt honored to see Tarangire in its infancy. The huge park seemed empty of all humans except the two of us bouncing along in a pickup truck alongside herds of elephants and giraffes. Tarangire did become, as Derek expected, a major tourist attraction.

Eventually we arrived at the airport in Dar es Salaam, where we parted. After a formal handshake, I said, "Thank you for going out of your way to see me off."

"Don't forget, you once piloted Jane and me in the park's boat from Kigoma to Gombe so she and I could relax and enjoy the ride together," he replied with a fatherly smile.

I smiled back, remembering Jane's smiles and his quiet affection on that boat ride. It was an interesting experience for a young man, to watch a relationship develop between two older mentors. Though I felt appreciated by both Jane and Derek, I later realized that I was also experiencing an inward journey—one that had started in college and accelerated at Gombe. I was discovering that I could be myself around people I cared about and receive their unconditional acceptance and support. I realize now that what I felt then was a growing sense of value. To feel valued for who you are is a transformative experience and one that I would carry forward into the next stage of my life.

After leaving Derek, I flew out on an evening plane to Amsterdam and then on to Minneapolis. The long plane rides allowed time for my thoughts to percolate—from feelings of accomplishment to a deep gratitude for the people and chimps I had learned from. As the plane approached the Twin Cities, I reached into my carry-on and pulled out a treasured gift.

Holding the Tanzanian figurine on my lap, I immediately pictured "Gombe Jane" (not the more hurried "Stanford Jane") walking toward me on the lakeshore just three days prior. Her eyes were sparkling and her gaze tranquil as she handed me the present, which she had just purchased in Dar es Salaam. "Your talisman represents a doctor," she had told me. The beautiful, dark-wood carving was of a man embracing a child. Now, sitting on the airplane in the dim light, I examined it, thinking that it embodied the healing comfort of touch and reassurance. Indeed, I felt strongly reassured at that moment.

Whenever I look at the familiar figurine now, I summon up the spirit of Gombe. I remember the excitement of a time when all of us student researchers were discovering more about the world and ourselves. We struggled to make career and other life decisions and leaned on each other just as the carved child leans into the arms of the doctor.

Today, I still recognize myself in that figurine. At times I am the doctor and at times the leaning child. I wonder if Jane ever views herself this way—Dr. Jane, as she is known to the Tanzanians and to the scientific community, with her doctorate from Cambridge. Perhaps she too has been the leaning child during challenging times in her passionate mission to study and protect the Gombe chimps and their forest home.

Over the next decades the chimpanzees would need Jane's protection, not only in Gombe but also all over the globe.

PART TWO

THE MEDICAL PATH

CHAPTER TEN

ENTERING THE MEDICAL FIELD

For three and a half decades after my return from the Gombe Stream Research Center, I concentrated on medical school, residency, and then work as a family physician at Group Health in Seattle. Yet I never forgot the chimps and always longed to return to Gombe. Daydreaming about Africa often kept me going through challenging times. My memories of Tanzania were reassuring and instructive. I found that I would often think of Fifi and Freud and their relationship as I practiced medicine. I remained close to the chimps, even though they were twelve thousand miles away.

Even during the grueling medical school years, I felt the need to share my unique experience, hoping this might inspire people to experience the great outdoors and focus on preserving wild species and lands in the United States and around the globe. Over the years, I gave talks with accompanying slide shows of the Gombe chimpanzees at high schools, medical gatherings, and even at the Museum of Natural History in Cleveland. I'm not sure the sophisticated museum donors were ready for my demonstration of pant-hoots and charging displays, but I probably woke up anyone who'd begun to fall asleep. I even hit celebrity status one day when I appeared on a morning talk show in Cleveland and returned to my medical school lectures to discover the entire class had watched the show as their first lecture of the day! Sharing my adventure with others became a way to keep the chimps fresh in my memory, even as more time elapsed. Now, forty years later, I recall not only the first time I saw Fifi termiting but also the laughter of the high school students in Cleveland watching films of Freud and Gremlin playing in the trees.

Fatherhood also brought both spontaneous joy and the stresses of parenting. My wife, Wendy, and I have two boys, Tommy and Patrick. Luckily for my sons, Wendy's talents made up for my deficiencies in music and sports other than soccer. As is true for so many American families, we became overly busy. Wendy was a high school biology teacher with many demands, and my life as a family doctor did not allow for a lot of unscheduled time either. Getting home late for dinner was not ideal for family life. My workdays were long, and sometimes I was up all night helping deliver babies, but my memories of Gombe continued to refresh my spirit when work and family life were intense.

The future of the chimpanzees and their endangered habitat also stayed on my mind over the years. In one recurring nightmare, I returned to

Gombe to discover that paved roads had displaced much of the forest, modern homes had been built in Kasekela Valley, and the chimps were confined to small, fenced-in areas. I awoke shaken by these images, until I realized it was only a dream—at least for now.

Since my first days as a physician, I've kept photos of Africa and the Gombe chimps on the walls of my exam rooms. These photos keep me in close touch with my adventure and give patients a glimpse of my days in Tanzania. In my physician bio, available to all new patients choosing a doctor, I list "cross-cultural interests" and "speaks Swahili." Because of these interests and my experiences abroad over the years, I have many patients whose roots are in Africa, Asia, South and Central America, and other parts of the world. I recently met a new patient who told my nurse that after reading my personal profile, he thought I was African. I was profoundly flattered.

One spring afternoon I walked into my office to catch my breath between patients. "This is too much," I said to myself, straining to stay on schedule and still devote enough time to each patient. Frustrated, I gazed out the window and caught a glimpse of sunlight dancing on the deep green leaves of a large maple across the street. With the shadows between the leaves changing shape in the breeze, I suddenly pictured three-year-old Freud swaying confidently in the branches as Fifi sat close by eating figs.

I then envisioned myself beneath the tree, contemplating their close relationship. I relaxed as I focused on this mother-infant pair communicating with each other through touch, gestures, and laughter. I imagined Melissa and her daughter Gremlin climbing up into the same tree to greet them. As I gazed out the window, the filtered sun warmed me, and I felt a familiar sense of joy and well-being.

"Dr. Crocker, your next patient is ready, and two more are waiting," my nurse told me gently to pull my head out of the trees. I turned away from the window, stepped into the gray-blue hallway with its fluorescent lighting, and headed to the exam room. I felt as if I had escaped to the forest, if only briefly, and been reenergized. I realized then how much I missed Gombe and nature. I wanted to be more connected to the natural rhythms of the day, to be less rushed, to breathe fresh air.

As a child, those connections were second nature. I had spent most of my free time outdoors in the park or the backyards of neighborhood friends. My mom would shoo us out after our homework and chores were done, calling us home again for meals.

In that moment in my office, I realized how important it still was for me to get outside into nature. I also started to accept that briefly visualizing myself in the Gombe forest or just stepping outside to go get a cup of coffee had to suffice for now.

In comparison, in the villages around Gombe, being outside is simply a fact of life. This can pose its own hardships, but also has its gifts. Villagers maintain a primal connection with nature—part of our species' evolutionary heritage. Modern life, however, has a way of rendering that heritage inconsequential, as I experienced with my own family. Over the years, home computers and other electronic devices increasingly began to draw all of us indoors.

When Pat was seven, I was invited to go on a three-day school field trip to the Olympic Peninsula as a chaperone and medical practitioner. As it was a nature-oriented trip, the students were not allowed to bring any kind of electronic device.

On the bus on the way, my son and his friend Josh complained about the policy. "I don't get why I can't bring my Game Boy," Josh said, frustrated. Josh was quite sedentary at home, I knew; he and Pat both loved Mario Bros., but it was clearly one of Josh's primary interests.

However, once we got there, we were too busy for electronics anyway. I watched Josh come alive with enthusiasm as we trekked through the rainforest, examining lichen, banana slugs, and tiny fish in the streams. No video games—but still, I was surrounded by smiling faces and laughter. We were all so much more at ease out among the trees, without the noise of modern city life.

Technology can also inhibit face-to-face, human-to-human communication and connection. Though it has led to invaluable advances in medical research and practice, there are also downsides. With the introduction of electronic medical records eight years ago, I can spend four or more hours a day practicing medicine by interacting with a computer screen. In response to this, I began to remind myself—as I do the families I see in my practice—to seek balance in life and to create room for experiencing nature.

Almost universally, patients I care for describe the joy and sense of healing when they're out in nature, whether in their own gardens, hiking in the mountains, or gazing at the evening stars. I'm convinced that a ninety-four-year-old patient of mine is alive and healthy because of his weekly trips to the Cascade Mountains to hike beside streams and wildflowers.

Given my experiences, I understand the current movement to "rewild" our lives. I understand why I long for more nature, more spontaneous movement of my body, and more time outdoors to feel the rhythm of the day, the changing seasons, the wind, the rain, the moon, and the sun. Our genetic wiring, which has not changed much over the past few million years, is better suited to a wilder existence than the very structured and stationary lifestyle we live. My time in the African forest connected me to a more nature-based life and enabled me to understand why I felt so alive and at home in the wilder world. I encourage my patients to reconnect with this side of themselves at every opportunity.

Even more significant to my medical practice was the image of Fifi and Freud making their way through the forest day after day, Freud prospering under the constant companionship and nurturing of his mother. Fifi's skills as a mother were embedded in my mind much more than the biochemistry formulas and cardiogram electrophysiology I learned in medical school. The latter were important, but not as spontaneously retrievable as the behavioral lessons I had learned at Gombe, and not as human.

One day early on in my practice, a patient named Ellen was talking to me about her two-year-old son, Sam, at his checkup. Ellen looked exhausted as the wiry blond toddler shredded the thin paper covering the exam table. When he climbed on a chair to reach for cupboard doors, she got up from her chair and whisked him back over to the exam table. He dangled over the edge and dropped back down to the floor, and we stopped talking and watched him drop Cheerios and grind them into the carpet with his feet and then go over to pull at the flexible arm of the exam light.

Ellen sighed deeply. "I'm sorry, Dr. Crocker."

Even though I did not yet have children of my own, I couldn't help but be reminded of the natural behavior of male chimps at this age. Sam was so much like Fifi's son, the three-year-old Freud, who would pull at palm fronds, swing through the air, wrestle with Gremlin, and toss sticks into the air. Of course Fifi didn't seem to worry about the mess of leaves and broken sticks Freud left on the forest floor. She would sit with Gremlin's mother, Melissa, and they'd peacefully groom each other as Freud carried on his exuberant play. In times of danger, such as when a male baboon approached, Fifi would grab him securely and guide him away, but she seemed to accept that Freud was going to be his curious self, whether climbing, swinging from dangling branches, or chasing other chimpanzees his own age.

Having witnessed Freud's antics, I felt comfortable listening to Ellen describe her feeling of a lack of control in her life, though I certainly empathized more after my first son turned two! But knowing how natural all these wild instincts were and why they may have evolved helped me feel better grounded in my role as a family doctor dealing with behavioral issues. Even just watching the *National Geographic* films of two- and three-year-old chimps playing tag with each other or wrestling on the ground highlighted the need for this rough-and-tumble play as a part of their early development. From their play, these little primates learn how to fit into their community and communicate with their peers.

I wonder, in fact, if suppressing too much of this lively exploring early in life might result in later problems such as depression, anxiety, or even aggression. Most humans have the skills later in life to temper their excitement and aggression, but young children don't. Though boundary setting is crucial at various stages of human development—and it varies with each child—from chimpanzees I learned the benefits of giving plenty of space for exploration and experimentation in the early years. During this time, children are using their senses to discover how things work. They do this by touching, feeling, and watching, and incorporating this learning into their real and make-believe worlds. Their brains make connections to help them prosper in a complex environment.

Though Ellen was well-read and progressive in her child-rearing, I wasn't quite ready to say to her, "You know, Sam really reminds me of a wild chimpanzee named Freud." Instead, I offered suggestions that seemed to help Ellen feel more accepting of her "wild" son based on my eight months of watching Fifi and Freud interact. What's true for chimps isn't always true for humans, but some of the basic, raw principles may apply. The idea of creating an "indestructible" and safe area in her house where Sam could let loose and explore in his own way was attractive to Ellen. I also said, "Just visualize Sam as an energetic, creative primate. He'll need redirection at times, yes, but also room for adventure and space to build a strong, coordinated body."

Because of my work with both chimps and human infants, I often see similarities between our species. Every age in a human child's life is important, but especially the early formative years. The bonds children form with their parents, grandparents, and guardians have a great impact on them. When Tommy was just five, I injured my spine playing soccer, and for two years didn't have the physical ability or energy to play sports with him. More important, I didn't have the focus to be as present and as interactive as I'd have liked. Though my wife filled in for much of what I couldn't do, I still look back at that time with regret because I wasn't able to maintain the close bond I'd had with Tommy prior to my injury.

With the chimp community, if the mother is injured or ill, there's usually no replacement. On occasion, however, an older sibling or unrelated adolescent or adult may be equipped with skills and available to help out. When Gremlin gave birth to her twins, Golden and Glitter, their older sibling, Gaia, helped her mother care for them.

In my practice I saw a young mother who was diagnosed with stage 4 colon cancer and had four young children to care for. During the last year of her life, she continued to be an attentive mom despite the physical pain of the cancer and the emotional pain of knowing that her children and husband were going to lose her.

"Deb has been amazingly involved with the kids, even from bed," her husband Dave informed me. "Sometimes I think she is a lot stronger than I would ever be, keeping up her parenting while dealing with the pain and nausea she is experiencing. She says I am the strong one, as I will now be a single parent without her." Such purpose in both chimps and humans seems to trump everything else in these successful mothers.

In the chimp community, just as with humans, the long period of infant dependency is crucial to developing the social and survival skills needed to succeed in their environment. Until the end, Fifi provided the right combination of support and playfulness to allow her offspring to become extremely competent adults. I can still picture her patiently waiting for Freud to practice his termiting skills. After snagging a large number of the tasty critters herself, Fifi seemed ready to move on, but she

A gathering of chimps socializing in a tree and surveying the valley. *Photo by John Crocker, 1973.*

Field assistant Hamisi Mkono guiding me through the valleys after we lost our chimps and began heading back to camp. *Photo courtesy of John Crocker, 1974.*

Hamisi Mkono and I climb a small tree high above the valley as we imitate our subjects. *Photo courtesy of John Crocker, 1974.*

ABOVE: Fifi with Frodo. *Photo copyright © The Jane Goodall Institute / by Jane Goodall, 1976.* BELOW: Young chimp swinging through the trees at Gombe. *Photo by Grant Heidrich, 1974.*

ABOVE: The Rift Mountains, part of the eastern border of the Great Rift Valley, separate human villages from the chimp communities. The chimps do not ascend to this barren territory but remain in the fertile valleys below, which slope down to Lake Tanganyika. As students, several of us slept out under the stars on the bare hill in the center of the photo. *Photo by Grant Heidrich, 1973.* BELOW: Melissa holding twins Gyre and Gimble. *Photo copyright © The Jane Goodall Institute, by Jane Goodall, 1978.*

Me spending a perilous night in an abandoned chimp nest built by adolescent Goblin. *Photo by Anthony Collins, 1973.*

ABOVE: Young baboon examining leaves in a tree at Gombe. *Photo by Grant Heidrich, 1974.* OPPOSITE TOP: Grant Heidrich relaxing on the beach at Gombe as one of his baboon subjects sits close by. *Photo courtesy of Grant Heinrich, 1974.* OPPOSITE BOTTOM: A paddler in a wooden boat at sunset brings produce back to his village from the town of Kigoma. *Photo by John Crocker, 1973.*

Hamisi Matama's family compound near Bubongo village. *Photo by John Crocker, 1973.*

stayed there and allowed Freud to continue poking his stick at a termite mound, trying to learn the complicated skill. In a year or two, he would be able to extract the high-protein food on his own. Luckily, nursing offspring for up to five years allows chimp mothers to give each infant the intimate time being close to her it needs to learn these crucial skills.

The obviously playful and joyful interactions of Flo and Fifi with their offspring seemed to result in their next generation's being socially comfortable adults. As with humans and the nurture-versus-nature debate, it might be difficult to tease out how much of this resultant behavior is due to genetics and how much to parenting styles, but even early studies of primates point out crucial elements such as close contact and support early on as at least partially contributing to later emotional health.

Flo faltered somewhat in the maternal care of her last two offspring, likely because of her advanced age and fragility, resulting in a very dependent youngster, Flint. When Flo's next offspring, Flame, died at an early age, Flint became even more dependent on Flo and seemed to regress to fill Flame's infantile position. Flint never progressed toward independence, as Flo was too weak to fend off his demands to sleep with her, nurse, and ride on her back, even when he was eight years old. His eventual death three weeks after his mother's death resulted from his depressed, emaciated state and illustrates the sophisticated and challenging parenting required of chimp mothers to ensure the emotional growth as well as the healthy physical development of their offspring. Fifi and Flo didn't have the support needed to help in the care of their offspring toward the end of their lives, even as they weakened. Because chimp mothers are very protective of their young, both Flo and Fifi likely would not have accepted another chimp's help with mothering. Their lives ended with an eight-year-old (for Flo) and two-year-old (for Fifi) still dependent on them physically and emotionally. Neither offspring was able to survive without his mother.

In my medical practice, I've seen impatience, neglect, and very tight controls in parenting situations. Sometimes I get a sense of mothering skills from descriptions of events at home. However, I also often get specific questions from parents that bring to mind Fifi's successful mothering in the wild, and Melissa's too.

Early in my practice, the mother of a five-month-old said, "My friend was saying that holding and rocking my daughter a lot might spoil her." I looked at her, and she was very lovingly holding the baby, but there was worry in her eyes. I thought back to my observations of the nursing and close body contact of young chimps and their mothers from sunrise to sunset, behaviors that extend for four to five years in the wild. This nurturing creates bonds that last a lifetime.

"Don't worry," I told the new mother. "You're forming a lifelong bond during these close times you share with your baby." She smiled and looked down at the infant in her arms. I added, "Enjoy this time! Hold her, feed her, rock her, and enjoy the closeness you share. Don't worry about spoiling her for another year—or more!"

That new mom also joined a parent support group in her neighborhood, which helped her immensely, especially having the alliance of other mothers who were going through similar struggles and joys with their infants. The chimpanzee females got to witness other mothers nurturing their young in the course of everyday life, but humans often need to more explicitly seek out support and information.

Some mothers and fathers in my practice have had difficulty nurturing their children due to stresses and distractions in their lives, such as working long hours outside the home, health problems like depression or anxiety, or simply not knowing how to parent after having had alcoholic or nonnurturing parents of their own. A distant or rigid form of parenting also seems to lead to major problems in children, who cannot adapt to that lack of warmth and nurturance along the way.

With all the complexities of modern society, social customs, and family expectations, I've found it instructive to pay close attention to nurturing, patience, and bonds between the primary caretaker—whether mom, dad, aunt, or nanny—and the children, as these were crucial parenting attributes in the chimp community I studied. Though Seattle and an East African forest are very different settings for raising young primates, I came to understand that some of the basic mothering behaviors of the chimps I'd observed made sense for human childhood development too.

CHAPTER ELEVEN

FOREST REFLECTIONS: JUNGLE INFLUENCES IN MY PRACTICE

What a privilege to be sitting next to a patient who trusts you enough to reveal their hidden fears and parts of their lives no one else might know about. The close connections I felt with my patients over the years kept me energized and enthused, and each patient encounter was like a new chapter in a novel.

Because of the adult male and female chimps I studied, I often tried to understand my patients' depression, anxiety, and aggressive behaviors from an evolutionary perspective. I would think about each person's actions and reactions in light of how evolution has hardwired us to survive. This has been crucial in my understanding and planning of treatment for patients with ADHD, anxiety, anger management issues, and stress-induced conditions such as chronic headaches and intestinal disturbances.

My time watching life and death among the chimps also prepared me to be an engaged listener. I could witness my patients as they struggled with the same joys and losses as any other species. When I watched my teenage patients lose the tight parental bonds formed in childhood and gradually enter the adult world, I remembered observing mirrored tension in the chimp community when Goblin and Pom began to seek their independence from their mothers. For example, Goblin would engage in tense interactions with alpha male Figan, followed by equally intense reassuring embraces with his mother, Melissa, clearly showing the emotional struggle churning within the preteen chimp. And Freud's and Gremlin's joy and laughter were not much different than the giggles and play wrestling of my sons and youngsters in my exam rooms. My time with the chimps trained me to look at the very basic needs that we humans have, such as reassurance through touch, having a purpose, and a connection to community, and not to underestimate the importance of those needs. I learned to slow down and offer my presence instead of a quick fix. In thinking about what is crucial for a healthy, developing forest chimp, I'm able to intuit what is missing in the lives of the children and adults I see every day in the office.

Some chimps and some people stand out in my mind as examples of how similar humans' basic needs, behaviors, and social structures are to those of wild chimpanzees. As I continually review what Gombe taught me, I keep case notes on certain patients. This helps me make new and ongoing connections between what I learned as a young researcher and my evolving work as a family practitioner.

Kumi and Community

In Gombe I forged lifelong ideas about community. Living and working with others in that environment showed me what it means to be relied upon as well as what it means to rely on others. Taking turns shopping for food, supporting each other during injury and illness, and watching out for one another during adventures deepened my sense of how vital social connections are to anyone's well-being. In family medicine, our training included assessing the social and psychological aspects of each patient in relation to the specific medical symptoms or chronic disease we are treating. A year after returning from Africa and starting medical school, I learned about the importance of these connections firsthand.

In the summer of 1975, with a $1,000 grant provided by Case Western School of Medicine, I joined three other medical students in a cross-cultural study of health beliefs and practices in the United States. I chose Japanese-American families in the San Francisco area and interviewed first- and second-generation families about their traditional and Western views of health care.

A Buddhist priest helped me find potential participants. I loved hearing family members describe their traditional remedies for various illnesses, such as drinking aloe vera mixtures for stomach problems and using moxibustion, a heat treatment, for back pain. Many also combined Western medical care with these more natural healing methods they learned growing up.

Notebook in hand, I arrived at a beautiful house in Marin County. Japanese gardens surrounded the house with bamboo trees and a trickling stream. When I knocked on the stately door, my study participant, Kumi, graciously welcomed me. After removing my shoes and entering the immaculate living room, I noticed the serenity of the home. Kumi lived with her Japanese-American husband, who was away working at the time of my visit. The couple had no children. Kumi was in her midthirties and dressed in a casual but elegant style.

Kumi was an issei, an immigrant who was born in Japan. I never ended up asking the questions I had intended to ask during our interview. Instead, I listened as Kumi spoke candidly about her feelings of isolation over the previous ten years, living away from family in Japan and having limited social connections in California.

"My situation is not the best here because I am too far away from my family," she confided.

Even though Kumi was a member of the local Buddhist temple, she felt she would be frowned upon if she complained about her situation and lack of friends, since she lived in a beautiful home and had no financial worries. Her main focus was taking care of her home and being a good companion to her mate. "My husband is very involved with his work, and my main purpose is to maintain the home, cook for the two of us, and provide for him emotionally."

Kumi poured out her feelings about her life. Growing up in Japan, she had been an excellent student and had considered becoming a doctor. She was also closely connected to her family. "I was fond of my mother and full of happiness when my cousins came to our house for celebrations. When I needed someone to talk with, I had two good friends from school I could turn to."

Kumi's self-discipline did not prevent her from having lots of fun with friends and relatives back home. Now, her support system, other than her husband, was thousands of miles away, and she could not see how it would ever be replaced in a California suburb.

As a medical student, I was probably not as intimidating as an experienced professional, which I believe made it easier for Kumi to reveal her feelings. I didn't have the training to know how to respond in a clinical manner, but I realized that Kumi wasn't really seeking medical advice, just reaching out and needing to talk.

Having so recently returned from Africa, I felt sad for Kumi as I recalled the comfort and support I received from the Gombe community. I too had been far away from family, but I had a purpose and a community of friends to provide emotional support. I thought also about the close-knit members of the various villages and how those villages functioned. It would be unlikely that people would feel isolated in such communities.

I never asked Kumi why she didn't have children, for fear that this might bring up another issue I wasn't qualified to address. I felt uneasy about inadvertently stirring up her emotions, and suggested she contact the Buddhist priest, who could provide spiritual support and perhaps find a role for her in the temple and a counselor to talk with. I hoped I'd been an empathetic listener, providing at least a small opening for Kumi to express some of her feelings. I realized more than ever the importance of viewing a person's health from many angles and how crucial it is for a strong extended family or community to provide emotional support in one's life. In Dan Buettner's book *Blue Zones*, five communities of people living in areas such as a coastal Italian island, Japan, and Loma Linda, California, were found to live to ripe old ages in part due to strong social bonds with family or community.

I wonder even today how Kumi dealt with the few choices she appeared to have. I wonder if, over the years, she found her way into social activities and groups that shared her interests and culture. I saw that for Kumi, feelings of loneliness were defining her day-to-day life. In recent years, there has been a lot of research on happiness and its effect on health. This research has found that the one factor that seems to assure well-being is not wealth, status, or marriage—but strong social bonds.

Carl and ADHD

Understanding the survival strategies of the chimps enabled me to be more engaged—and less intimidated—by certain conditions I encountered later in my practice. This was certainly true when I worked with an eight-year-old named Carl, who had ADHD, or attention deficit hyperactivity disorder, which manifests itself in focus problems and in being restless. During an appointment with Carl and his mother, I viewed his condition from an evolutionary perspective.

Even when I spoke directly to Carl, his eyes searched the exam room and his body was in constant motion. As he tapped his hand on the exam table and kept crossing and uncrossing his legs, he resembled, in a way, a caged, restless chimp.

Many youngsters with ADD or ADHD excel in sports; many have superb endurance and high intelligence. They often have a powerful focus in certain areas, such as computer work and activities requiring concentration, such as engineering. Often, however, they have profound difficulties in school because of the need to sit still in the classroom and listen quietly. Their performance drops and so does their self-esteem. With lower self-esteem comes a steady loss of confidence in all areas of their lives—even activities they once felt secure doing become fraught with anxiety and fear of failure.

Both Carl and his mother lit up instantly when I asked his mother, "Can Carl stay focused when he's left alone to build with Legos?"

Carl nodded with interest, and his mother said, "Oh my gosh, you wouldn't believe what Carl can do with Legos!" She described his ability to focus and follow directions to build large and complex structures while Carl looked proud.

Working with Legos requires concentration, motor skills, and willingness to persevere. There are other examples of "play" activities that reveal a child's talents, despite seeming like they're just recreational outlets. Playing with Legos illustrated some of Carl's strengths, and as we discussed what Lego play required of Carl, his mother saw his engineering talents and creativity validated. These traits were in full swing during this activity.

His mother said, smiling broadly, "I thought working with Legos was just playtime!" This activity hadn't communicated anything significant to her about her son, but now both Carl and his mother saw its importance.

I asked Carl, "How's school going?" and he just shrugged. He wanted to talk about his Power Rangers and his new *Star Wars* laser sword. He took a startling leap off the table to demonstrate a powerful action-figure pose, holding his light saber.

As he did this, I told Carl, "I would definitely choose you to be my guide if I were traveling in an African forest and facing a threatening leopard," and he perked up even more, holding his chest a little prouder. I could see that he would be physically and emotionally driven to defend against an enemy, his searching eyes and quick reflexes crucial to spotting and responding to danger. Carl was not overly aggressive; he just needed to move. His brain didn't have the pathways activated to allow him to be sedentary and calm—it was not in his genetic wiring. The usual confines of modern living were challenging to him.

I couldn't help but think that Frodo would have scored high on the ADHD scale. He exhibited many ADHD traits, such as restlessness, high activity level, searching eyes, and an inability to patiently attend to subtle social cues from others, yet he successfully served in his community. He was an excellent hunter and attained alpha male status, channeling his innate aggressive and athletic tendencies to the benefit of the group as a whole. Though most male chimpanzees show aggressive behavior during dramatic displays or when they perceive a threat, Frodo wore those attributes on his chest most of his waking hours. With his large, muscular body, he demonstrated a genetic predisposition to a high level of aggression when he hunted with superb success, nearly broke Jane's neck, and stole an infant from a Tanzanian mother.

Carl's genes evolved for a reason, even though the standard school setting was not the ideal place for him to exhibit his talents. The evolutionary perspective seemed to shed some needed light on Carl's clinical condition. My reinforcement of his personal strengths and talents helped him to feel better about himself and attain feelings of confidence. With that came trust—from both Carl and his mother.

In Carl's case, the medication Ritalin was needed to help him focus in a school setting. It helped him manage surroundings that didn't always play to his strengths. He was able to function and succeed in a school environment without undue struggle and frustration. While such medications must be managed carefully, they can activate a pathway in the brain for the type of focus required in the classroom. When a child does indeed need such medications, and the medications work the way they're supposed to, the child is able to adapt to the demands of school and of peer relationships. Being able to adapt and feel successful leads to a much more positive view of one's self and being an overall happier person.

My heart would always hold a special spot for kids who were struggling with ADHD. My professional understanding of how talented these patients can be has grown steadily over the course of my career. Hard-wiring of the brain resulting in attention disorders also likely bundles with it wiring for many positive characteristics. These traits often go unnoticed because of our focus on the less desirable ones. An impulsive child is often creative, but only the impulsiveness catches the adult eye. In reviewing the literature about common positive traits in people with ADHD, I have seen consistent statements about heightened creativity, spontaneity, intuitiveness, and thinking outside of the box. In my practice, I coach parents to be mindful about focusing on their children's strengths, likely improving the children's self-esteem and even their social interactions with peers and teachers.

In the chimp community, both alpha male aggression and ADHD characteristics such as those exhibited by Frodo are necessary to the species' survival. Though the survival of our own species may now depend on collaboration and intelligence more than on powerful aggression, we can still be hardwired to be more like Frodo than is useful. But when we understand that Frodo-like hardwiring, we're able to view and use those traits in more productive ways. By recognizing these primitive, instinctive behaviors and reactions within ourselves, we can channel them appropriately. Sports, wilderness adventures, and scientific and technological exploration may continue to provide adaptive outlets for our innate aggressive or restless tendencies.

Sandra and a Sense of Purpose in Chimps and Humans

My observations of chimpanzees convinced me that they are pro-grammed to have a strong purpose—whether it's caring for an infant, patrolling the border of the community, or simply searching for food. In humans, this might correlate with a deep inner need or drive to be useful, feel appreciated, or feel a sense of accomplishment.

When I encounter patients with depression or anxiety, there are usually complex reasons for the condition. This often includes a family history of depression or difficult early-childhood experiences, but I also take into account the role of purpose in people's lives. I recognize how evolutionary genetics may program us to feel rewarded for accomplishing tasks we set for ourselves. If a person loses that sense of purpose, depression or anxiety can follow. Sometimes, patients may not even be aware that their sense of purpose has been altered for some reason. They may not make a connection between their feelings of depression and their loss of purpose.

Some of the happiest retirees in my practice talk about the activi-ties that fill their lives now that they no longer work at traditional jobs. They may care for their grandkids, volunteer at senior centers, or work part-time to supplement a fixed income. These activities increase retirees' feelings of self-worth, giving their lives enhanced meaning as they adapt to aging and the changes it brings. Active and involved retirees tend to be more at peace with the passage of time and their changing place in the world. They don't feel isolated or shunted aside.

In contrast, I also see male and female patients who fall into depression after they've retired from their careers. Their strong sense of identity from their careers has been suddenly lost. Even though a retiree may have time to relax and have fun, there is still a major adjustment to losing this strong life purpose. As one sixty-eight-year-old man told me, "I never thought I would miss work so much."

In a chimp community in the wild, purpose is ever present. There is no "retirement" for chimpanzees. Foraging for food, fending off baboons or other groups of chimps, building elaborate nests each night, and caring for infants—mother chimps give birth into their forties—are required duties that might also make any primate feel fulfilled.

As a doctor, I see firsthand the key role a sense of purpose can play in a patient's feeling of well-being. I entered the examination room one morning to find my patient Sandra there, looking miserable. She had two young children, and a husband who traveled a lot.

Sandra admitted, "I guess I've been struggling with depression off and on for a while. I don't know why—nothing's really wrong. You'd think it would be better when Mark was home, but I almost feel like it's worse then."

She went on to explain that when her husband was away on business, she put on her running shoes, literally, and managed to keep it all together. Her husband's absences put added pressure on her but also provided her a kind of outlet. Running the household and immersing herself in added responsibilities seemed to either distract her from her depression or alleviate it temporarily. With her husband away, Sandra felt a greater sense of purpose, which filled up the time and prevented her from engaging in introspection. Nonetheless, her depression would always come back. Being endlessly busy can provide a kind of solace, but it won't solve the underlying problem if there is more than just purpose involved.

I suggested that Sandra talk to a counselor to help her understand more about the roots of her depression. "You need to make sure your own needs are being met so you can manage the daily challenges of your busy life." I thought that with a counselor's help she might also consider the possibility that her husband's absences freed her of certain conflicts related to him and that she needed to work on those issues.

When I next saw Sandra, she said, "Thanks for sending me to that counselor. We talked and I realized that I had a conflict between wanting a career in teaching, like my mother had, and wanting to be available for my kids." She looked much more animated than she had on her last visit as she filled me in. The counselor had said Sandra needed to find her clarity of purpose and identify what she really wanted. Also, she needed to take into consideration the effects meeting her needs would have on her children and husband.

By identifying what she wanted to do with her life, Sandra was able to find ways to follow that path without sacrificing other purposeful demands. She began taking night classes so she could teach a few years down the line, once her children entered school. She found a way to balance their needs with her own. Her face shone with enthusiasm as she shared her vision for the next phase of her life as a working mom. She had set a goal and found new purpose.

With such a seemingly high rate of depression in my medical practice, I decided to ask Jane about the incidence of depression in wild chimpanzees. I hadn't recognized it in the chimps I had studied during my student days at Gombe. After pondering the question for a few seconds, she said, "Genetically inherited depression would be selected out of any given chimp population's gene pool in the wild." Typical symptoms of depression, including lack of interest, social isolation, low energy or fatigue, and lack of appetite would interfere with the chimp's ability to survive in the wild. Situational depression however, did occur in the Gombe chimps, especially when a young chimp lost his or her mother.

Flint was deeply depressed when his mother died, but lasting cases of depression other than those related to a death were not seen in the Gombe chimps. In the wild, they generally thrive emotionally in an environment they're wired for, fulfilling what we think of as purpose.

For some humans, a lack of fulfillment can be insurmountable and lead to anxiety and depression, drug abuse, or all three. Additionally, feeling conflicted about one's own desires can produce anxiety and depression. Modern human life can be filled with foggy uncertainties. We are faced with more choices than chimps and therefore have a stronger need to define our roles and desires in life. The chimps seem to live a life of clarity and drive, as survival behaviors consume most of their waking hours, at least in adults.

Frances and Eddie: Life's Changing Seasons

There were certainly spiritual moments during my time as a student at Gombe, and I experienced several epiphanies while living in the wild with our primate cousins. One day, four months after arriving as a student, I found myself next to screaming chimps soon after a colobus kill—but I recognized that my heart-pounding response to danger was offset by having an experienced field assistant next to me as well as by my growing understanding of chimp aggression. The forest was my classroom, and the caring attitude of my fellow researchers helped me realize the importance of bonding with those around me. This would be a lifelong influence, and it inspired me as a physician. It sparked an enduring drive to connect with my patients on a very deep and long-term level.

I formed an exceptionally strong bond with Frances, a seventy-five-year-old who had been a patient of mine for twenty-five years. I had enormous admiration for this down-to-earth, wiry, and gracious environmental activist who had given much of her time and money to help preserve the beautiful lands of Washington State.

One day when I walked into the exam room to see Frances, she burst into tears. This unshakable, physically strong, emotionally poised hero of mine was more vulnerable than I'd ever seen her before. "I'm sorry," she said, taking a tissue.

I shook my head. "Don't be sorry. What's going on, Frances?"

"I'm terrified Eddie's Alzheimer's disease is advancing rapidly." She managed to choke out her story. Frances was miserable, feeling she was losing her relationship with her husband after fifty years. Once she had managed to share that, she sobbed for several minutes, then looked up at me with a tentative smile and said, "Damn, life can be tough."

I knew at that moment that despite her pain, Frances did not want to give up hope. The hardy soul who was spending her retirement years repairing trails and fund-raising to preserve Mother Earth was determined to help keep her relationship with her husband alive. She still had the fiery will that had seen her through the many environmental fights she waged. I found her indomitable spirit remarkable.

Because I'd known Frances for so long and we shared a common bond with our love of nature, I didn't immediately start talking about the specifics of Alzheimer's, or talk about the effectiveness of medication that can be mildly helpful in some patients with dementia. Instead, we sat together and talked about life. In the small exam room decorated with photos of my own past adventures, I sat close to Frances while she took me back to her earlier days with Eddie. As we talked, it felt as if we were both reliving some of her past experiences with him and also mourning together. We talked about when she first met her husband.

"I boarded a city bus in the pouring rain wearing the most ridiculous old hat to keep my head dry," Frances told me, her eyes bright with the memory. "Eddie was twenty-seven years old and very handsome. He looked over at me and asked, 'Where did you ever find a classy hat like that?' When I took it off, the pooled rain on the top poured onto his lap. We both laughed so hard that I'm sure people on the bus assumed we were the best of friends, which we soon became. We had forty-five more years of friendship during our marriage, and I hope we have a few more."

This was not a medical encounter, and neither was it a diagnostic exchange. This was a shared emotional and spiritual experience similar to ones I'd had in the forest. In Gombe, more than anywhere else, I had witnessed the coming and going of life, the changing seasons, and the rising and setting sun. My long history with Frances certainly made this intimate sharing possible, but I think it could have happened even between strangers if both were tuned in to their feelings and the spirit of the natural world.

Counselors point out that when most people hear a person express emotional pain and distress over a particular situation, they respond by trying to fix it. Many times, however, the person experiencing the pain needs someone to just listen and be with them. An attempt to "fix" things may not provide comfort. There are experiences in life that aren't "fixable." Loss remains loss; sorrow remains sorrow. These are human experiences that we all face at one time or another. When we extend comfort and understanding, the challenge we face is often to simply "be" with the sufferer. As soon as we try to "make it better," we're no longer present with him or her in the journey—we've removed ourselves. In this situation with Frances, I knew we could discuss treatment options for Eddie later, and we did. For the moment, all she needed was to talk and be heard. Since so much of my time studying the chimps involved merely observing them, this may have engrained in me the habit of quietly observing and listening to my patients before jumping in and trying to fix their problems.

As Eddie's condition worsened over time, Frances adapted surprisingly well. She discovered other outlets for her emotional well-being in an Alzheimer's support group and enjoyed the support of others in her community. She has accepted that her love for Eddie endures but that his disease has forever changed her communication with him. Now when I see her, we talk more about her than about Eddie, as Frances enters a new season in her life's journey.

Robert: Nature Eases Anxiety

During thirty years practicing medicine, I've seen patients with anxiety of some form nearly every day. At Gombe, I saw how anxiety could be adaptive but also detrimental if sustained for long periods. For a chimp to patrol the border of his community in a relaxed state might not produce the desired outcome, but to remain in a constantly vigilant and tense state of fight-or-flight is also not sustainable. The high-ranking chimps with pent-up energy and anxiety would let it all out when needed, in displays or just quick drumbeats on a hollow tree buttress. Here, again, my jungle experiences illuminated my practice.

Robert was a patient in his thirties. Single, he worked full-time in the maritime business and enjoyed his work and social life, but his usual level of anxiety had increased. We explored several possibilities and could find no clear answer as to why he felt more anxious than usual.

Soon his increased anxiety was affecting even the most mundane aspects of his life. Robert spoke nervously during one of our appointments and then got a bit choked up as he told me, "I don't even like to drive the car anymore because I'm afraid something will happen to me."

I pictured the male chimps Satan, Figan, and Evered pacing along the border that separates their community from the southern community. Always in motion, they seemed tense, but they appeared to be directing the adrenaline that naturally built up in their systems into powerful swaggers and movements. The chimps released their pent-up hormones in aggressive displays and charging behavior.

Then I pictured Robert sitting in traffic with his anxiety and tense muscles, worried and seemingly paralyzed. No movement or release of tension was possible, except for perhaps groaning or shifting in his seat. The adrenaline would likely be passively circulating through his blood as the byproducts and cortisol built up, making Robert feel uneasy and eventually tired. If only he could charge down a hillside, climb high into a tree, or let out a loud vocalization that people could hear hundreds of yards away, he might well feel better.

There are drugs that work well in relieving chronic anxiety and are usually safe and non-habit-forming, including some that are also used to treat depression by elevating the level of the neurotransmitter serotonin. Robert wasn't interested in taking medication, so we designed a rigorous exercise program that involved jogging every day. It could be as little as fifteen minutes, but twenty to thirty were even better. I did not bring up tree climbing or hurling kerosene cans.

For anxious patients, I sometimes demonstrate relaxation techniques that they can do on their own. One that I shared with Robert involves slow breathing and picturing oneself on a warm, sandy beach or in another favorite sanctuary. These visualization techniques have been used in meditation and prayer for centuries and can also be used to self-calm. Patients find that they can create a sense of tranquility during this exercise and are then able to transfer it to other parts of their lives. Eventually, they can get to the relaxed state with just a quick visualization or a few deep breaths.

During the guided breathing exercise, Robert seemed to enter a relaxed state right away. He closed his eyes and sank deeper into the chair; his arms and legs didn't move. He looked so restful that I stopped talking and just let him be. Occasionally patients fall sound asleep in a few minutes. When I asked Robert to slowly open his eyes, he smiled and didn't say anything. I asked, "Do you want to describe your peaceful place?"

He looked at me with raised eyebrows and said, "It's not a warm, sandy beach. It's the apple orchard at my grandparents' farm in Eastern Washington. My grandparents always took time for me, and whenever I visited them, they would walk with me through the orchard. They loved telling me stories and hearing about my life. I knew I could tell them anything. I always felt secure when I was around them. At times, I sense the same comfort even when I'm alone among the apple trees."

"Wow!" I said. "That sounds like a wonderful place."

I thought of my own daydreams of the African forest, and I could relate to the security Robert's grandparents had provided him amid the trees. His apple orchard was like my Gombe. We all carry some special place inside us that we can retreat to during times of stress. To find that place, and hang on to it, can give us refuge when life's stress becomes too much. And it's so often outdoors, and it's so often a place in which we've experienced loving support.

Robert had received the training to do his professional work well but not the tools he needed to accommodate his specific wiring of being a highly vigilant and cautious person. In the chimp community, touching while grooming and embracing after excitement can create calmness in seconds. Vigorous travel through the forest also likely helps expend energy and release the tension built up during everyday chimp life. In our modern and fast-moving human society, however, long periods of time can pass before we touch another person, breathe fresh air, or move our bodies, other than our hands to manipulate our electronics. Interpersonal conflicts with friends and coworkers can cause stress and anxiety that remain bottled up for long stretches of time. I wonder sometimes if our complex brains that enabled us to build a computer and fly to the moon have also pushed us into undesirable emotional states.

Perhaps as humans, our next mission should be to better understand how to create that feeling of calm and those joyful interpersonal connections that nurture us throughout our lifetimes. I suppose this was one of the reasons I became so excited to study early primate development, searching for clues as to what we need early on in our lives to support long-lasting inner peace and self-confidence. With each passing year in my practice of medicine, I have learned more about how an individual's unique genetic makeup will influence this formula. For instance, a more extroverted and less sensitive person might have thrived in Robert's job and social situation without the accompanying paralyzing anxiety and without needing to learn the coping skills Robert had to learn to be more happy and well-adjusted.

Ruth, Margie, Scott, Faben, and Madame Bee: Creative Adaptations in Humans and Chimps

Having studied human biology in college and then observing chimps as they adapted to tough conditions, I often think about how people and other primates learn to face environments for which they might not be genetically or emotionally suited. Sometimes I even picture certain friends or patients in an entirely different place and time, such as a mountain village in Nepal, a town in northern Italy in the 1800s, or an ancient African forest, thinking that their traits might be more appropriate there. Carl, for instance, might have been a shining star in a past century helping out on the farm using his mechanical skills or caring for animals, maintaining a strong sense of purpose. He might have been greatly appreciated instead of criticized. Robert, with his anxiety, might have been a good castle guard in medieval times as he paced the perimeter watching for invaders. In addition to some of my patients and the Gombe chimps, my grandmother inspired me with her creative adaptation skills.

My grandmother Ruth, who lived to 106, surrounded herself with beautiful trinkets and sparkling jewelry and always wore elaborately designed scarves to complement her attire. She dreamed of having a stately Tudor home like the ones her husband designed and built overlooking their city, but my grandparents' house in Portland was small and on a busy street, with Grandfather's business office attached to the front on the lower level, and the upper level rented out as an apartment. Nonetheless, Grandma Ruth made the living room and dining room on the main floor her sanctuary. Ceramic Chinese lanterns above the fireplace, oriental carpets, and fancy satin curtains made you feel like you were in a grand palace. She adapted to her modest circumstances by creating a vibrant, luxurious space that expressed her artistic nature.

Grandma Ruth had also purchased an ornate, gold-colored, tinseled staff. Somehow this decorative item made its way from her house to mine when I was growing up. At times, my siblings and I would flash it around in play. It became a reminder to me of Grandma Ruth's sumptuous living room—emblematic of her personal fantasy—but more important, it became part of my own escapes into fantasy and daydreams, as I imagined the staff belonging to a prince or a queen. Imagining these different scenarios served as an escape from conflicts at home or boring lessons in the classroom. Even today, I can picture the staff and immediately feel transported into creative times in my life, whether daydreams about flying carpets or dressing in wild costumes for Halloween.

In my practice, a teenage patient, Margie, also used fantasy to adapt to daily life. When I met Margie, she initially appeared to be a nonconversant, disinterested, and unanimated teenager. She wore mostly black.

Then I asked about the book she was reading, which had winged dragons, towering castles, and swirling capes on its cover. With a smile that could light up a dark room, Margie said, "This is what keeps me sane." It was post–Harry Potter reading that consumed much of her time.

Fantasy-adventure books may have been just a straightforward interest, but Margie was both very smart and quite sensitive to her surroundings. She admitted being overwhelmed by a fast-paced family life with both parents in business and keeping tight schedules at home. Margie couldn't see the point of this pace, and wanted more free time just to relax and be herself. She was expected to get top grades and do well in swimming, and was encouraged to run for student body vice president for the leadership experience. Instead, Margie found her creative niche by writing for the school paper and cultivating a few close, mildly eccentric friends who made her social life interesting. She had identified a safe and stimulating environment suited to her personality and talents. Reading fantasy novels had helped her balance the demands that her world placed on her by propelling her into creative outlets that were meaningful to her. It was intriguing to hear how her individualistic behavior helped her adapt to her stressful surroundings.

It has also been my privilege as a doctor to learn how individuals adapt and succeed while facing extreme physical challenges. Scott was a patient who had become quadriplegic in his twenties when riding in the back of a pickup truck that struck another car. Today, forty-two years after his accident, Scott is still thriving with his wife and grown step-children. Though dependent on help from others and struggling with his health, he maintains an amazing attitude. His deep-rooted optimism and humor about life serve him well. He always communicates with me from a positive perspective, even when shaking from a fever or acknowl-edging the need for a leg amputation. "It won't affect how I dance in my wheelchair, which I've done for the past twenty years," he told me before his leg surgery. I'm in awe at how Scott and other patients forge ahead with their lives despite the obstacles, dispelling preconceived assumptions about how their lives will evolve.

I remember observing two chimps at Gombe who survived and adapted to physical challenges. Madame Bee and Faben contracted polio, and each lost the use of an arm, which would seem insurmountable in the wild. Nevertheless, Faben adapted to his handicap by learning to walk upright, and with one very strong, functional arm, Madame Bee raised all her offspring. Since survival in the chimp community requires that each individual care for himself or herself after childhood, watching these chimps adapt and live without help was very impressive.

We parents all try hard to engender adaptability in our children. Still, a child's temperament or innate traits, which we can't control, can deeply affect the equation. The shy, sensitive child or adult will certainly learn to adapt differently than the outspoken extrovert. In the case of the chimps, it's clear that their need to survive in the wild tests their temperament to a high degree, as in the case of losing a mother. Flint, at age eight, was the only Gombe infant over the age of six to die as a result of his mother's death. Perhaps his emotional dependence on Flo was partly due to his dependent and submissive nature—in contrast to the more independent chimps, who could adapt to surrogate mothers as they survived the loss of their own.

I believe that one's ability to bond well and be adaptive in close relationships with other people may be a result of forming a trusting bond early in life with one or both parents, but I realize that some of my patients learned to adapt well to life's challenges despite a harsh early environment. Perhaps genetic traits are at play in these cases, once again underscoring the importance of both nature and nurture in our development.

Whenever I ponder the remarkable adaptive behaviors of humans and chimps, my thoughts return to Grandma Ruth. Though her gold-tinseled staff is long gone, I keep it in my memory, conjuring images of designing floats for the Portland Rose Festival Junior Parade and of Grandma creating an exotic wardrobe and adorning her modest home with treasures she'd found at thrift shops or yard sales.

But most important, I'm reminded of Grandma Ruth's mantra: "I am very grateful for having had such a wonderful and long life." I try to remember the word "grateful" when life is challenging. I hope my sons can use their creativity to adapt to life's trials and that Fifi's great-great-grand-offspring can do the same. And I hope that being raised by loving parents helps them too.

Personal Stories Keep Me Going

Many other patient relationships have kept my interest in and appreciation for my profession going strong over the years. This compensates for the necessary but more tedious time I spend on the computer viewing lab results and nursing-home requests and refilling prescriptions for medications. In some ways, my months of observing chimpanzees without ever getting bored remind me of the thousands of patient encounters that still bring enthusiasm and joy to my work.

My eldest son, Tommy, with his interest in Eastern religion and meditation, pointed out to me that being totally "present" during these patient interactions might also explain why I feel so content when I'm in the exam room with them. I can't think ahead to the next consultation or commitment or look back, regretting not having completed some task or another. When I'm with a patient there is no opportunity for my mind to wander. I take in a variety of information that requires uninterrupted focus. I must listen carefully to the patient's history and symptoms as I formulate a diagnosis, contemplate a possible treatment plan and consider psycho-social issues that might be contributing. It is rather like being a detective vigilantly sifting through information for clues. Some refer to this state of intense focus as being "in the zone." I am there, fully absorbed, mindfully attentive with the patient just as I was while observing the chimpanzees at Gombe.

Also, I think about another gift I've received as a physician, in addition to the privilege of being a part of my patients' lives. I learned that what's beneficial for the patient may be equally so for the physician. Early in my practice I scheduled weekly appointments at the end of the day to accommodate a college football player named Jake, who was having panic attacks and experiencing severe anxiety and wasn't performing well in class or in his sport. Although I prescribed non-habit-forming medication to help Jake function during the day, I also taught him relaxation and stress-reduction techniques as tools to help manage and eventually reduce his panic episodes. The appointments followed a full day's work and ended past the dinner hour (I was single at the time), but I realized that I always felt better on those days, perhaps because I too was benefiting from the techniques we practiced together.

Over decades of caring for patients at the same medical center, some of the patients I delivered were now asking me to deliver their children, I guess making me a "grand-doctor." I was happy—my practice and my family were my life—but as I entered my twenty-eighth year of doctoring, I became aware of the need to physically reconnect with my past jungle experience, to see and feel the elements of the Gombe forest. I longed to see the chimps again. Work was fulfilling yet exhausting, and family life was fun but also demanding. Though it seemed impossible to pull away, I knew I couldn't wait much longer.

PART THREE

RETURN TO GOMBE

CHAPTER TWELVE

GOMBE CALLING

S adly, I didn't stay in touch with Hamisi. After the first year, during
which we wrote each other two or three letters in Swahili, I stopped
corresponding. I didn't feel good about this, but life was busy and it
seemed difficult to keep up a long-distance friendship, especially in
Swahili. I thought about Hamisi often, wondered what his life was like,
and more than once made tentative plans for a trip to Gombe, but for a
thousand reasons—once even after I had purchased an airline ticket—
those plans were never realized.

Then, after thirty-six years of practicing medicine and helping raise
my family, my daydream of returning became reality. It was 2009. I was
fifty-eight years old. I had begun writing about Gombe, and even my
family could tell that when I sat down at the computer I was excited and
happy to immerse myself in my African memories.

It was my older son, Tommy, ever my inspiration, who broke through what had been wishful thinking on my part by saying, "Dad, why not go back to Tanzania and finish writing there?"

At age twenty, Tommy was now engaging as a young adult, having shed most of his adolescent aloofness with his parents. Like me he had a shy edge, but unlike me he could play high-level soccer with finesse and had also performed in high school musicals. He was beginning to adjust to his small-town college setting at Colgate University in snowy upstate New York after having attended a big, diverse public high school in Seattle.

I went blank for a few seconds, then turned to him and replied, "Great idea! Why don't you come with me?"

To my surprise, Tommy looked impassive, and then reluctant when I suggested he come with me to Africa. I had thought he would jump at the chance to go—what young person doesn't want an adventure? I used all the tactics I could think of to persuade him to come along, enticing him with visions of the two of us watching the chimps build nests and hiking to Bubongo Village. I loved the idea of introducing Gombe to my son. I had been Tommy's age, in my early twenties, when I was a student researcher there.

But something was holding Tommy back, and I realized that I didn't want to pressure him. It wouldn't be fun to take a dream trip with a hesitant companion, and it wouldn't be right for him either. Though Tommy remained cautious and didn't show much excitement about the trip, he came and found me at my computer three weeks later, and said, "OK, Dad, I'll go. I want to."

Patrick, at age ten, would likely have jumped at the offer to come along as he was already skilled at making authentic-sounding pant-hoots, which he demonstrated during some of my presentations at his school. But chimps can be aggressive toward smaller humans and Gombe has strict rules preventing those under fifteen years of age from visiting.

As I began to make travel plans, I discovered that securing local plane reservations and guides in Tanzania was hit-or-miss. I often did not hear back from those I attempted to contact, and I was still uncertain about whether we would connect with people I'd known and some of the chimps I'd studied.

We would fly from Seattle to Arusha, Tanzania, and visit the vast Ngorongoro Crater. Next we would fly to Tanzania's largest city, Dar es Salaam, and possibly see Jane, who knew we might be there when she was passing through overnight, then fly eight hundred miles west to Kigoma to catch a park boat to Gombe.

Though I'm usually more spontaneous with travel, I wanted to make a detailed plan to get the most out of the two and a half weeks I was able to be away from work. This was a big father-son event in our lives. Though we had no idea what we would see and do at each juncture, what mattered most was that we would experience this unique pilgrimage together.

In the weeks before we left, I thought about the dangers that could arise while we were tracking wild chimpanzees. Jane had reported brutal behavior among some of the male chimps after I left in 1974, and I wondered about the possibility of our being viciously attacked by aggressive chimps, which was rare but not unheard of at Gombe. Being far from civilization, I was also concerned about exposure to malaria and other diseases and infections, even though we would take all the necessary precautions. Compared to my first trip as a student, when I was more carefree, especially when it came to health issues, I now felt the extra responsibility of having my son with me. This time I was going as a father, and it was as a father that I would experience every moment of this trip.

Deep inside, I realized that I was also concerned about being disappointed. I wasn't sure if I would still feel at home in a place where I'd had some of the most intense experiences of my life. Unfamiliar park staff would greet me; the landscape might have changed; and the young chimps I studied in the 1970s would now be in their golden years. How would I react if it just didn't seem like the Gombe I'd once known? Finally, I worried that Tommy, not having the same emotional investment, might wish he had stayed home with his friends. I wanted my son to love Gombe as I had, but I couldn't force that to happen—Tommy's experience would be his own.

All those worries aside, it was clear to me that I wanted to make this journey with Tommy. To be with my son on a return to Gombe was especially meaningful, since Tommy represented the next generation, the young men and women who would become the future stewards of the natural world. A strong internal voice was saying, "I have to do this!"

As for Tommy's concerns, he finally voiced his fears on our plane ride to Africa, confiding, "Dad, I'm not sure I can go very far into the jungle to see the chimps. It's the poisonous snakes—the black mambas." I turned to look at his face, which was full of emotion: fear, but also a desire to please, and maybe some shame.

He continued, "I heard that black mambas roam the Gombe valleys where we'll be hiking and crawling around after the chimps." He was terrified of them. I suddenly realized that my son's initial reluctance about joining me wasn't from the embarrassment a young guy might feel about spending weeks with his father, but instead from true anxiety and fear.

I tried to practice what I always preached—listening and patience. After Tommy was done talking, I said, "You can choose not to go into the forest to see the chimps and still experience the other parts of Gombe that are incredibly beautiful and interesting." *Just not as interesting as the chimps*, I thought, my heart falling a bit. But my feelings couldn't dictate how Tommy might feel. I thought for a bit. I didn't want to dismiss his worries, but also didn't want to encourage his fears. So all I said further was, "You might have the chance to understand your fears better when we get to the camp."

That opportunity arose sooner than either of us could have imagined. Just before we left Arusha, our first stop in Africa, our curiosity was aroused by a sign that read SNAKE PARK. We hesitated a moment and then decided to tour the park. When we reached the black mamba exhibit, Tommy stopped. We read that a black mamba can travel as fast as fourteen miles per hour as it chases its prey, keeping its head and the first third of its body poised above the ground. The longest venomous snake in Africa, it averages eight feet in length, and some reach fourteen feet long. Mamba venom contains neurotoxins, and the mortality rate is close to 100 percent unless the antivenin is given immediately. Also known as the "seven-step snake"—you only have seven steps left after being bitten—the black mamba preys on birds, rats, small chickens, and bush babies; its main predator is the mongoose. Humans are sometimes bitten if they inadvertently corner the snakes or startle them.

As I moved ahead to the next exhibit, Tommy stayed with the caged mamba. He watched the snake closely through the glass as it watched him. The stare-down lasted several minutes. Later that day, Tommy proudly announced, "I figured out a way to alleviate my fear of the mambas. I envisioned myself today as a snake warrior attacking the mamba with some primitive tool just before it could strike me." The strong visual image of himself with power and determination allowed him to venture into the jungle days later without such paralyzing fear. His exposure to the snake in a controlled environment helped him imagine it in an unpredictable environment. He also understood more about mambas, which gave him a greater sense of calm.

When I was Tommy's age in Gombe, I was just too naive to understand the risks of being around this snake, or perhaps I was overly confident because no one at the camp had ever been bitten. Now, as Tommy's fear of mambas waned, my previously muted fears intensified. I had to wonder, on many levels, what was I getting myself—and my son—into by venturing back to this captivating but unpredictable and sometimes dangerous land.

CHAPTER THIRTEEN

REUNION WITH JANE

Mostly for Tommy's sake, I planned to start our African adventure in a more natural setting than the big city of Dar es Salaam. Consequently, we landed in Arusha and spent a day in the Ngorongoro Crater, where more than twenty-five thousand wild animals roam. The crater lies close to the Olduvai Gorge, the steep-sided ravine in the Great Rift Valley where Louis and Mary Leakey discovered fossils of early humans from nearly two million years ago, and where Jane Goodall joined the couple for her first work in Africa.

When we laid eyes on the crater, I thought, *There is a sense of romance here.* The stark beauty of the acacia trees, the brilliantly colored birds, and the vast savanna filled with lions and herds of giraffe were nearly unbelievable. In our lodge room the first night, we heard the hooves of buffalo, and spotted elephants hiking up out of the crater in the moon-light, to Tommy's delight.

As expected, Ngorongoro contrasted sharply with our arrival the next day in Dar. To get to our hotel near Jane's house, we took a three-hour cab ride from the airport through congested downtown streets in rush-hour traffic with numerous clunky trucks spewing sooty exhaust into the humid air. Pedestrians seemed to be constantly dodging between the vehicles on the road. We finally made it to the hotel and I realized my neck muscles were as tight as boards. Despite all our efforts, there was still a good chance we would spend the night in our hotel room and leave the next day without seeing Jane. I had a strong desire to see Jane on this "homecoming" trip, but didn't want to get my hopes up, since she had warned me that her plans could change at a moment's notice. She would be passing through Tanzania during the second week of July, so we had a small window of opportunity to meet up with her.

Both Tommy and I relaxed when we got to the hotel and Tony, the Scotsman in charge of the baboon study and my past Gombe comrade, walked right up to us. Wow! A thrill shot through my spine as his kind smile and familiar manner brought me immediately back to Gombe. I had always felt safe and relaxed around him. Lots of hugs and smiles ensued. Tommy loved meeting him, and Tony commented how similar Tommy was to the younger me at Gombe. The three of us drove on narrow, unpaved streets with mud-and-brick houses and small markets on our way to Jane's house, north of the city. We arrived just before 10:00 P.M. Though Jane had visited our family several times through the years, I felt tremendously nervous.

Jane's modest wooden house, located on the shores of the Indian Ocean and surrounded by mango trees, hadn't changed much since I had last visited. As we walked up to the house, the smell of the salty beach, mango blossoms, and burning candles hurled me back thirty-six years to when Hamisi and I had stayed here on our way to climb Kilimanjaro.

Knowing Jane would likely be in deep thought or in conversation with someone, we quietly walked through the open entryway door and then headed toward the kitchen, where we heard voices. When I peeked in, I didn't recognize the three people there, so we headed to the living room. Tommy gave me a nod as he spotted Jane talking to a few friends and staff in the screened dining room. We walked slowly toward her.

Jane, wearing a sweater and casual cotton pants, continued talking but acknowledged us with a smile. I noticed I felt a little shaky as we approached. She opened the circle of people and announced in an incredibly warm voice, "John and his son Tommy are here to join us." She greeted me with a hug and received my kiss on each cheek.

"It's great to see you, Jane, especially here in Tanzania," I said excitedly.

Tommy, looking enthusiastic, was very gracious as Jane said, "It's nice you could come with your father on his trip back to Gombe."

So began a two-hour reunion with Jane, Grub, Tony, and three people who worked for the Jane Goodall Institute in Tanzania. Sitting around a large wooden table, we sipped whiskey and shared stories, memories, and updates on our lives. The sea breeze coming through the screens and the dim lighting reminded me of evenings at Gombe back in 1973 with Jane, Tony, and other researchers, talking at the end of the day near the beach on Lake Tanganyika. I wasn't sure how much Tommy was getting out of all the stories, but he took it all in, seemed quite relaxed, and drank whiskey for the first time in his life.

The conversation turned to the Gombe chimps, and Tony described the developmental milestones of twin chimpanzees Golden and Glitter. I enjoyed hearing about these successful offspring of Gremlin, whom I had studied at Gombe for eight months.

Tony recounted one of their recent adventures: "Golden and Glitter were in the treetops feeding on fruit when a cloud of termites flew by. They both stood up on the springy branches and began to snatch the tasty insects out of the sky and devour them. They looked like they were waving to the world. A cinematographer caught great footage of their acrobatic performance."

Jane was thrilled, laughing and clapping as Tony reenacted the scene by waving his arms and pretending to eat the morsels he was reaching for. Despite her travel fatigue, she carried on as though nothing could stop her from enjoying the evening, especially hearing the latest news about Gombe.

Turning to Tommy, she asked, "What are your interests at college?"

"I've decided not to pursue medicine," my son explained. "I really like philosophy." Then he added with some pride, "I enjoy being a resident dorm counselor."

Jane listened to him closely, her bright eyes fixed on him.

At that moment, I wished I were more like my son—authentic and natural. Instead of always worrying about what to say, Tommy was simply being himself. He was totally present, whereas I felt happy but anxious, wanting everything to go smoothly, as if that were up to me. I remembered my dad displaying responsibility in social situations and how he would insert his directives to gain command and feel in control. Here I was doing the same, at least in my thoughts. *Stop it,* I told myself. *Learn from your son. Let the evening be the way the evening chooses to be. Be patient. Be present.*

Jane talked about her memories of Tommy growing up, as she had seen him at various times over the years, and also of my younger son, Patrick. "I remember the pillow fight I had with Patrick on my stop in Seattle," she said, smiling. "How is he doing? And how is Wendy? It's a shame they couldn't come too."

I appreciated Jane asking about my family and remembering details of past events in our lives. Finally, I felt myself relax as the conversation turned to other people in the room. Perhaps the whiskey was also contributing, but I realized it didn't really matter what Jane or Tommy or I said from that point on. We were together. There was a simple communal joy from just being in the same room, watching the evening pass, hearing the ocean waves, and remembering our times together. Tommy, Jane, and I were reunited once again, this time in Africa, on our individual journeys in life at ages twenty, fifty-eight, and seventy-five. Indeed, one April a few years earlier, I had wished Jane happy birthday and she quipped, "I will always be seventeen years older than you." But to me she would always remain ageless.

I loved talking with Jane's son, who was now a father. People still affectionately called him Grub in place of his more formal name, Hugo Eric Louis. I had not seen him since I left Gombe when he was only seven years old. This evening I observed him sitting quietly in a chair close by, and he smiled as I approached. He had a calm and inviting presence and was very comfortable with himself. With his rugged Dutch/English looks and excellent physical condition, Grub looked younger than his forty-three years.

I began telling Grub about a day I spent with him at Gombe in 1973. "We were on the beach when you showed me how the pods from a particular bush would burst open when heated by the sun, scattering the seeds yards away. I remember you telling me, 'You pick off the pods when they're dry but not yet open and place them in the hot sun.' You then grabbed my hand to take me to the perfect large rock to place them on." I assumed he wouldn't remember the day in question, but in fact he went on to describe even more details.

"I think we also explored the water life together in the shallow water at the lake's edge."

"Oh yeah, we did!" I exclaimed. I distinctly remembered his fearless diving into the lake and our underwater exploration.

Our conversation turned to Grub's interests in fishing and archeology. "I hope to explore the Tendaguru Beds in southeast Tanzania," he explained.

In fact, he and Jane would be leaving at six o'clock the next morning with a German film crew to make a documentary about the area. The Tendaguru Beds are very rich in dinosaur fossils from the late Jurassic period, 150 million years ago. A German mining engineer discovered the beds in 1906. The area is also home to the many elephants, lions, and buffalo found in game parks in other areas of Tanzania. Jane has established a fund to help preserve the area by forming partnerships with the local people to aid them economically while preserving the wildlife.

Grub said that he wasn't sure it had been the best choice to pursue fishing and lobster catching as an occupation in his twenties and thirties, knowing it conflicted with Jane's mission of protecting all forms of animal life. "It may not have been wise to get involved in the fishing industry, but at the time it was what I knew and enjoyed most," he said.

I could definitely see both Jane and Hugo Van Lawick in their son's eyes and expressions. Grub's eyes were the color and shape of his father's but he had his mother's welcoming gaze. I wondered how he viewed his place as the only heir in the legendary Van Lawick-Goodall family, and tried to picture what he might be doing ten years from now—perhaps archeology, oceanography, or animal conservation in Africa. I sensed he would contribute significantly to Tanzanian society in his own way. His beautiful children's mother was Tanzanian, and Grub had lived in Tanzania his whole life. Most recently, he had been designing and building a unique type of boat for fishing and other uses near Dar es Salaam.

As midnight approached, I noticed Jane looking more tired, so I decided it was time for us to leave. She was so comfortable with groups of people that she might have stayed up even later just to be gracious, but I didn't want to take any more of her time. I could tell by her smile as she walked Tony, Tommy, and me to the door that she was happy to have had some of her Gombe family around her that evening. As was usual when I spent time with Jane, I left with more hope in my soul about the world and more confidence in myself.

On the ride back to our hotel, where we would pack for our trip to Kigoma in the morning, I glanced at Tommy. *He looks a lot like me*, I thought, and I realized both how long and how short thirty-six years was—the time between meeting Jane when I was his age and visiting her now. Jane was a grandmother now. Time had passed so very quickly. Along with a tinge of melancholy about aging, I felt the tender satisfaction of watching my son grow into a caring young man.

In some ways, our trip to Gombe after Tommy's sophomore year of college was well timed. He seemed to enjoy more in-depth conversation and being "present" at a time when I seemed to be less patient and was feeling more pressure trying to keep up with my practice and family demands. Tommy was learning to enjoy the journey while I was working hard to get to the next destination. Courses in Eastern religion and philosophy may have guided Tommy in a more mindful direction while my work schedule led me to be more task-oriented. Our trip together allowed me to see this contrast and think of ways to get some of the "journey" back into my life. Also, being in Tanzania, where a considerable amount of my early adult growth had taken place, may have enhanced my desire for more spontaneity and thoughtful moments.

I was discovering new ways to connect with Tommy. I no longer worried that I could name only five players in the NBA or that I didn't know who the quarterback for the Green Bay Packers was. A new father-son relationship was forming, and I was learning as much—if not more—from Tommy as he was from me. I was gaining insight into the importance of thoughtful living, and he had seen me push through difficult parts of my life with hard work and optimism. He made a heartwarming comment one night as we unpacked in a hotel room: "Dad, your subtle reminders of the importance of keeping my room neat paid off—I actually like having it more organized now."

I started noticing also, as we traveled and talked together, that Tommy preferred to hear more about my past struggles than my accomplishments. At the same time, he was revealing his own insecurities to me—not for advice, but just to share. In a quiet but confident voice he told me, "I've noticed on this trip that when I talk to you about things that make me anxious, you don't get all stirred up and worried like you and Mom used to do when I lived at home. You mostly just listen. It makes it easier to share things with you." At that moment I felt a private happiness that our father-son bond, which I felt had been interrupted by a disabling and painful sports injury I suffered when he was five, and then again during his aloof adolescence, was being renewed. It might not have happened had we not launched off on this reunion together. What had seemed a reconnection with the chimps, Jane, and Gombe was also turning out to be an unexpected reconnection with my eldest son.

CHAPTER FOURTEEN
GOMBE RETURN

As Tommy and I flew on a prop plane from Dar es Salaam to Kigoma, I anxiously anticipated the changes I might encounter. Though I would not see Fifi, as she had died six years previously, there was a chance I might see Freud or perhaps Gremlin. I had studied both of them when they were two years old, and now they were in their late thirties. I had mourned Fifi, that caring chimp I came to know so well, when I heard she was gone, but I was curious to see the legacy of her influence in Freud and Frodo, whom I knew were alive.

Our eight-hundred-mile flight took us near Mount Kilimanjaro. "Dad," Tommy said, pointing, "I thought Kilimanjaro had a lot more snow on the top, from the pictures I saw in *National Geographic*."

"Oh my God," I whispered in shock as I looked out the window.

It was unsettling to see the severe lack of snow on the mountain compared to my trip thirty-six years earlier. Now we know from scientists' studies of the mountain's glaciers that one particular large ice field has lost 50 percent of its mass since 2000 from climate change and possibly from local factors such as deforestation at its base. At this rate of melting, there would no more snow on the mountain by 2018!

We then flew over the Serengeti—the vast grasslands that support extraordinary migrations—and on to the town of Kigoma, on the shores of Lake Tanganyika. Tommy laughed as the plane landed at the airport, just a small dirt runway with grass and a few basic structures.

As we stepped out of the plane, I had a sudden feeling of dread. Although I had sent a letter three weeks previously to the park director regarding our plans, I had received no reply. Our phones were of no use here. We knew no one. *What would our next step be?* Our only option was to follow the other passengers across the reddish-brown dirt pathway. Then to our utter surprise, at the exit gate stood a rugged, distinguished-looking man, Lameck, the park director for Gombe, holding a sign with CROCKER written in big letters. I looked at Tommy who could tell I had been worried and we both smiled and picked up our pace to greet this savior. Although my Swahili was okay, he spoke perfect English, and with his relaxed manner and non-stop smile, we could tell we were in good hands. We loaded our luggage and climbed into the sturdy park truck. Lameck brought us home with him for a lunch prepared by his wife. I immediately felt welcomed, diving into our Tanzanian-style lunch of chicken and cassava, speaking English with our hosts and relaxing in their cozy home.

Then, taking charge of our journey, Lameck drove us to a small beach where park rangers were waiting for us. Lameck, the boat driver, and the rest of us set out in the blue wooden park boat for the three-hour trip north on the vast lake. The warm breeze felt good as we talked the whole time with the park rangers, Tommy smiling as he conversed in

English with a young ranger. The quaint fishing villages looked similar to those I recalled from nearly four decades earlier, which reassured me that the beautiful lakeside was still quite natural.

As we approached Gombe, my heart rate accelerated. In my mind I had rehearsed this moment of return. I pictured the familiar faces of field assistants smiling and welcoming us to the camp, just as they had decades ago when I arrived as a student. Tommy could sense my excitement, and his eyes eagerly scanned the rugged shore and the fertile valley leading up to the Rift Mountains, taking everything in.

But as the boat got closer, nothing looked familiar. Instead of the expansive beach that I remembered, I found that dense, leafy trees had advanced toward the lake, leaving only ten feet of beach. A small green concrete arch printed with GOMBE STREAM NATIONAL PARK and a picture of a chimp face became my focus as the boat approached the shore. The arch looked slightly commercial and I wished it weren't there. Tommy may have heard me mutter, "I don't like that," under my breath. Equally shocking, a two-story concrete building loomed in the vegetation where formerly a more rustic, thatched meetinghouse had stood, the location of our research team's nightly meals and gatherings. Expressing my uncertainty, I turned to Lameck and asked, "Is this Gombe?"

He laughed and replied, "Yes, this is Gombe."

I must admit that I felt disappointed. I've learned over time to conceal dismay and sadness as an adaptive strategy. As children, my siblings and I felt that our dad always wanted us to be happy, and there seemed to be an incentive to look that way to avoid disappointing him. In medicine, sometimes for the good of the patient I needed to subdue sadness and communicate confidence, support, and empathy when breaking bad news to patients.

I remember telling Steve, a thirty-nine-year-old patient of mine, "Steve, it is quite likely that you have a lymphoma," after feeling a mass in his upper abdomen and receiving the ultrasound report I had ordered.

I tried to look into his hopeful but worried brown eyes and sound confident and encouraging while feeling torn apart inside that this young man had a potentially fatal condition. This never gets easier, but as a doctor you do develop coping strategies. I found over the years that approaching this issue requires a slowness with myself—I can't rush my words to my patient, and I must try to communicate a sense of balanced reason. Observing a patient closely as I convey bad news is helpful because I can assess how best to work with that person going forward; it also shows me what his or her thresholds are. Patients or patients' families sometimes told me how important this time was for sensing a doctor's compassion and caring spirit. Conveying those things can require that doctors conceal their own inner reactions of dismay and loss, and communicate only calm understanding and empathy.

Though wanting to conceal my initial disappointment about how Gombe had changed pales in comparison to delivering bad medical news, my medical experience allowed me to maintain a smiling facade as we reached shore. As the boat was secured and we stepped onto the beach, I recalled something Grandma Ruth used to say to me when I was young: "Never go down to the ocean with a notion of what you will find." Her wise words reminded me that the only constant in life is change. And my son, who was nearly four decades younger than me, was there and seeing Gombe with fresh eyes. I didn't want my feelings to color his first impressions.

As I stepped out of the boat onto the sandy beach, I heard one of the staff say, "There's Hamisi."

I looked down the beach and saw a figure walking toward us—and I recognized Hamisi Matama, my dear friend and primary field guide when I was a student researcher. Hearing that I was coming, Hamisi had hiked the five hours from his village and waited most of the afternoon to greet me on my return. After thirty-six years, there he was walking toward me, a big smile on his face. My disappointment turned to joy as

I took off down the beach. Hugging Hamisi, I knew that I had at last returned to the Gombe I remembered.

Even now, in his early fifties, Hamisi had very erect posture and looked very strong. He wore a perfectly fitting white *taqiyah* (a Muslim skullcap), khaki pants, and a plaid shirt. As we stood there together, vivid memories came flooding back: climbing Kilimanjaro with Hamisi and watching his astonished face as he touched snow for the first time; hiking together to his village to meet his family; and hours trekking through the forest to observe the chimps—they all came alive. A deep pleasure coursed through me as I realized that our friendship had lasted through decades of separation.

As we walked back to the boat, laughter rose from some of the Gombe staff there, who were excited about our reunion. Tommy had been waiting and talking with them to give Hamisi and me time to get reacquainted. Reaching the boat, I turned and introduced Hamisi to Tommy. My son extended his hand to Hamisi, and they each said, "*Jambo*," hi. Hamisi looked more serious than I had expected, and I couldn't tell if this was cultural or just shyness on his part. Later, when Hamisi and I had time for a more intimate conversation, I realized that meeting Tommy might have reminded Hamisi of one of his own sons, who had died at age eighteen several years earlier.

Introductions made, two Tanzanian staff members I didn't recognize accompanied Tommy and me to check in to our simple accommodations in the concrete building that also housed the dining hall above us. The structure was separated from the beach by a row of leafy trees that provided welcome shade. Hamisi waited near the beach for us.

I was tired and hot as we entered our room, and I felt Lake Tanganyika beckoning to me with its clear water that had invigorated me each day during my stay here in 1973. Peering out the window at the waves lapping the shore, I told Tommy, "I have to jump in the lake."

Tommy was more patient and practical, however. During his twenty years of life, he had developed a strong social intelligence and was good at picking up subtle cues from people about their needs. Being a center midfielder in soccer, he could also make quick decisions about where the team needed to move the ball. In this case, he had realized that Hamisi had a plan of his own, which I hadn't fully recognized. Tommy politely redirected me by saying, "Dad, we can swim anytime. Let's go with Hamisi."

We walked back to the beach, where Hamisi pulled out an old black-and-white photo of the two of us taken in 1973—Hamisi, age seventeen, and me, age twenty-two. The photo, saved for nearly four decades, was mildewed and wrinkled, but the clarity of our faces was amazing.

I allowed myself to really absorb the moment. The fact that Hamisi had saved the picture and carried it with him to show me made me feel even closer to him. I put my hand on his shoulder and he immediately put his on mine. I looked at him and declared in Swahili, "Good friends."

My choice to follow Tommy's instinctive lead here was a good one, especially since Hamisi seemed very proud to have organized a get-together. He had been planning to take us to the men's quarters down the beach to meet other field assistants, treat us to cold sodas, and hang out and talk. We all headed to the field assistants' camp fifty yards south. When we arrived at the area, which I knew well, we met some of the newer field assistants and a few older ones I had known from my student days, but none I knew well.

They welcomed us into a small outdoor enclosure attached to one of the temporary homes where they stayed while working at Gombe. Ten to fifteen people could live at this location at one time. The screened-in areas protected them from the baboons that roamed the beach and sometimes searched for food near their quarters. While we were in there, a baboon walked by and observed us in the "caged" area. I laughed; it was the opposite of a zoo.

One shocking change since my first trip here was that there was a large-screen TV in the enclosure. It seemed so out of place but was obviously something the field assistants loved to watch after following chimps all day. Attentive eyes were glued to the screen when I arrived, but the assistants soon turned it off as we began talking. Mostly news and sports events seemed to capture their attention. There had been no electronic devices at Gombe when I was a student. In response to my surprise, Hamisi said, "A bigger generator supplies electricity to the camp to power things like the TV and the dining hall lights."

Tommy and I had brought many laminated copies of the photographs I took in 1973 to share with people at the camp. The field assistants hadn't owned cameras back then, so they smiled and exclaimed over seeing the old images of themselves or their relatives. One of the younger field assistants pointed to a picture and said, "That is my father!" He stared and stared at the photograph, touching it and smiling periodically. Tommy remarked later that it had been very moving to witness the man's joy at seeing his father at a young age. We all sat around laughing, talking, and telling stories about the pictures, which I gave them to keep.

After our visit with the field assistants, Tommy and I decided it was the right moment to give Hamisi the gift we had brought for him: a wristwatch he had asked for thirty-five years earlier, a year after I left Gombe. I still recall the letter a friend translated into English for me, which gave all the details of the requested watch, including the type of band, the color, the stop and start timer, and a date display. I'd never made the purchase back then, because I was busy with medical school, because I wasn't sure how to send it with the best chance of getting it all the way to his village, and because of my concern that the other field assistants would feel left out. I always thought I should have sent it though, and I had carried the guilt with me for decades.

It was almost a relief when Hamisi mentioned needing a watch in response to a message I'd sent before this trip, asking if he had a special request for a small item I could bring. I asked Tommy to buy a nice one;

he found a very attractive gold-colored watch with a flexible wristband, date display, and timer, and packed it carefully in a small box.

When I handed Hamisi the elegant box containing the watch, he looked at it and smiled. After carefully opening the box and seeing the watch shine in the afternoon sun, Hamisi placed it on his wrist. He wore it the entire time we were with him, despite the band being way too big for his wrist. He looked magnificently proud to be wearing it. Tommy kept smiling and looking at me when he saw the appreciation on the face of my dear friend.

"*Asante sana*," thank you very much, Hamisi said softly again and again as he stared at it.

We finally did have time to plunge into Lake Tanganyika before dinner. Tommy had been concerned about the water cobras I'd seen while swimming when I was a student here, but none had been spotted recently. There is also a parasitic disease called bilharzia that is carried by snails in other parts of the lake, but there was none at Gombe, in part because of Jane's environmental work. Sanitation and erosion-prevention measures pioneered by her organization, Lake Tanganyika Catchment Reforestation and Education (TACARE, or "Take Care"), likely helped prevent this area of the lake from becoming contaminated. The disease, caused by parasites called schistosomes, can be contracted while wading or swimming in freshwater lakes and ponds in many parts of Africa, but at Gombe, the clear waters were safe to explore.

The dark green-blue water looked irresistible just as the sun was setting, and we dove into the cool depths. As I rose back up to the surface, I thought it felt like a ceremonial baptism for Tommy and a renewal for me, returning to this beautiful habitat. I floated on my back and swam out from shore to gaze at the mountains above Gombe.

That evening, two Tanzanian cooks prepared a dinner of mushroom soup, fresh fish, and local fruit for the eight visitors at the camp, including Tommy and me and several tourists from England

and Germany. It was now official: I was a tourist. I definitely did not eat so well during my time as a student, though our meals had been wholesome and very tasty. This meal seemed gourmet in comparison because of the special sauces and spices used in the preparation. In the past, we typically ate a dish with beans, bananas, and palm-nut oil or fried degas fish.

At dinner, the camp manager said, "This is Abdul. He's the field guide who will take you out to see the chimps tomorrow." The man he pointed to, who seemed to be in his midtwenties, nodded and smiled warmly. Abdul was very articulate, and he spoke English quite well. We took an instant liking to him. Much of the money he made as a field guide he used to support his ailing mother, who lived in Kigoma. Tommy and I connected with Abdul right away because of his sincerity, his comfort with speaking English, and his friendly nature as he described our meeting place for the next morning. We could not have dreamed of a better guide to lead us through the forest.

As I was finishing my last bites of fish, Hamisi told me, "I'm returning to my village early tomorrow, so I should wish you farewell." He walked us down to our room, and we said our good-byes. I was disappointed to see him leave, but we were planning to visit him and his family in just a few days. I said, "We are so excited about our upcoming visit!" and he gave me a gentle smile.

Our room in the concrete building near the beach offered two comfortable beds and running water. When I had lived in my old thatched hut high up in the forest, I would lay awake sometimes, listening to cicadas and branches blowing in the breeze. Here in our modern room, I heard only the soft murmur of conversations from people in the dining area directly above our room. I wished we could have stayed in my forest hut or in Jane's original rustic house on the beach, which was now available to travelers, but the staff had assumed we would enjoy the amenities of the more modern building.

I had wanted Tommy to experience the nighttime sounds of the African forest. But battling my disappointment, I reminded myself, *I'm sure he doesn't care—in fact, he probably feels more secure in this solid structure, protected from reptiles, baboons, and other potentially problematic visitors.* Right before falling asleep, I smiled with deep contentment, knowing that my son and I would be tracking chimps together the next day, and knowing that my dear friend Hamisi was still here in this extraordinary African forest—and now, so was I.

CHAPTER FIFTEEN

BACK TO THE FOREST

A forceful pounding filled my chest, and my nervous stomach did flip-flops. After three hours of arduous and discouraging search for the chimps high in Kasekela Valley, we had suddenly come across a group of seven chimps, both males and females, right on the path. My jaw dropped when Abdul said, "There is Freud."

I had followed this powerful chimp's every move for so much of my eight-month stay in Gombe. He and Fifi, his mother, had been the main focus of my mother-infant fieldwork. Now, thirty-six years later, the thirty-nine-year-old chimp strode past me. He had become alpha male as predicted, but was past his prime and no longer *bwana mkubwa*, big man in the community.

Freud had a relaxed and confident manner befitting his heritage. I caught a close-up of him before he began grooming with three other adult chimps twenty feet from us, and I was happy to see that some of his gestures and facial features looked familiar. Then a deep sadness filled me; I realized I had never seen him without his mother. Fifi was gone. She had disappeared four years earlier and was presumed dead. Although I knew she had died, I had really hoped to see her again.

The chimps came within ten feet of us, and we slowly backed away and sat quietly in the brush to avoid disturbing them. Wide-eyed, Tommy shot me a smile. I leaned over and whispered excitedly to him, "I can't believe we're sitting here with Freud. I never thought I would be introducing you to him."

"Amazing to see him so close up," Tommy quietly replied. "I feel like I already know Freud from all your stories about him and Fifi." Then he just beamed for the next several minutes. Looking at him, I thought, *He's part of the story now.*

The chimps seemed to totally ignore our small group, including a few field assistants who had just joined the three of us. We sat very quietly and observed them before they moved off quickly, heading north. We did not follow. Researchers had been restricted from going beyond a certain point in the northern part of the park because of a northern group of chimps in that region that were not habituated to humans. Unlike the Kasekela chimps, they were dangerous because they hadn't yet come to trust human observers.

"One of those chimps attacked a villager, so we are not allowed to enter that area," Abdul explained. "We can wait here because our chimps never go far into the northern chimps' territory and will likely return on this same trail." We waited patiently for two hours, talking and sampling the juicy yellow flesh of the *mbula* fruits that had dropped to the ground—and then, sure enough, Freud and the others in the group returned. We took off following them again.

Suddenly, a very different energy developed within the group. In sharp contrast to the quiet grooming and feeding, chimp screaming and calls of excitement started echoing through the trees. Tommy looked confused. One of the field assistants in our group shouted, "Frodo has killed a baby colobus monkey!" Another field assistant who was fifty yards ahead of us had radioed him the news. I had a sudden flashback: thirty-six years earlier, before there were radios at the camp, we would alert each other by producing a high-pitched staccato war whoop that resonated across the valley.

I remembered that there was always enormous excitement after a kill as the chimps gathered around begging for meat and aggressively fending off other animals. The frightening sounds grew louder.

Tommy asked, "Shouldn't we head in the opposite direction?"

I smiled confidently and replied, "We'll be fine. We're not a threat. Just don't ask Frodo for a piece of the meat." He chuckled.

We all hurried to the site of the kill and I immediately saw Frodo high in a tree. He was feasting on the baby monkey and sharing some of the meat with female chimps and a nearby male. I didn't see Freud, but he might have been hanging out on the periphery, and there was so much commotion that I may have missed him. We heard every type of vocal call chimps use to communicate: pant-hoots, screaming, grunting, and panting sounds as they interacted and shared in the kill.

Standing protected in the foliage below, watching all the activity and commotion, Tommy whispered, "I can't believe how fierce some of them are!"

"Those are fear grins on the chimps crouching and reaching out to Frodo for scraps of meat," I whispered to Tommy as he stared at the scene, wide-eyed.

Many of the other field assistants and a few researchers gathered to observe the chimps' behavior after the kill. We ducked under some low branches so as to be less noticeable as the screaming continued in the trees above.

It was then that I think Tommy realized for the first time that we were truly in a privileged situation. It was something I had understood as a student decades ago, when I was able to watch the lively interactions of the Gombe community of chimps as a special guest. I sensed that Tommy understood that this was a rare and remarkable experience, being physically close to strong, wild animals, capable of sudden killing, and yet we knew that they tolerated our presence and would likely not harm us. Tommy showed intense interest and no more fear of the powerful chimps than most visitors do.

After watching more of the chimps' excitement, Abdul led us away and we finished the afternoon by hiking along a streambed. My muscles were tired; I was no longer accustomed to scrambling up and down the valley through the dense foliage. When I caught sight of the beach, I gratefully assumed that our day of trekking and observing had come to an end, and I started eyeing the water.

After forty-five minutes of resting on the beach and talking, however, Abdul said, "Are you ready to find the chimps and watch them build their nests for the evening?"

"*Ndiyo,*" yes, I told him, reluctantly abandoning the idea of a plunge in the lake and instead stretching my legs for another trek high into the valley.

The forest was definitely quieter now, except for the many birds chirping and calling. Though most visitors to Gombe are required to return to the beach camp by five o'clock to ensure all visitors and field assistants are accounted for before dark, because of my experience as a student, Tommy and I were allowed to stay with the chimps until sunset, around seven thirty.

Abdul worked hard to try to locate a group of chimps so we could watch them build their nests. We finally found a group around six, but after feeding for an hour they suddenly took off and rapidly moved deep into a thicket, making it impossible for us to traverse the rugged terrain and witness their nest building. The disappointment was overwhelming.

And yet the hour prior to their exit had left me with one of the deepest feelings of connection to the generations of chimps at Gombe. Tommy and I were able to observe Frodo, who was part of this group, lounging on a grassy hillside. As the sun settled on the horizon, we watched him sprawled out in the long grass, munching on dark orange fruits. We were also sprawled out, no more than twenty-five feet away from the legendary chimp, peering at him through the dry foliage. Knowing that he had been the largest and most powerful alpha male but was now in "retirement," we both thought he looked very much at peace. At that moment it was hard to imagine we were looking at the same Frodo who had once been perhaps the most skillful hunter and fiercest leader of the Gombe chimpanzees.

Frodo ascended to alpha male status in 1997. He was huge, with an unusually strong physique, and according to Jane, had overthrown his older brother, Freud, with brute strength and ruled the community with an iron fist. Frodo once attacked a primatologist who was observing him, and even pounced on Jane when he was in his twenties, nearly breaking her neck—an unusual occurrence, since the chimps were accustomed to her presence and had not been aggressive toward her before.

In 2002, Frodo snatched a fourteen-month-old human girl from her mother. The mother was passing through Gombe National Park with her niece, who had the baby wrapped onto her back, the way many Tanzanian women carry their infants. Frodo grabbed the infant, took her up into a tree, and killed her. As gruesome a tragedy as this was, to Frodo the child probably looked like a baby colobus monkey, which was among his natural prey and thus occasionally on the menu. He was a very good hunter and probably did not distinguish between species. None of this helps console a mother in the unbearable loss of a child; rather, it was a terrible warning to keep infants and young children far away from the territory the chimps call home.

Several months after this incident, Frodo contracted a parasitic illness that weakened him temporarily. During that time he lost his place as head male and over time became more complacent. There was no declared leader of the Gombe community for two years until Sheldon became bwana mkubwa. Abdul, Tommy, and I knew that Frodo hadn't lost his skill as a hunter, however, as we had witnessed the aftermath of his killing a colobus monkey earlier in the day.

The scene we now watched on the hillside reflected a milder side of Frodo. "He looks so self-satisfied right now," Tommy commented. "He looks like he needs nothing else in life except the fresh air and milk apples. Maybe we're all working too hard back home."

"I agree," I said, thinking about my long work hours.

I felt very connected to Frodo, being so close to him, and having known his brother and mother so well. When Frodo was in his twenties, the adolescent males would watch him closely and observe his ferocious displays and confidence within the chimp community. As I watched him now, I wondered if I could have projected more confidence—leaving out the ferocious displays—as a father figure to my two sons.

My father was more like Frodo in his conduct—high energy and definitely in charge, both at home and at work. I responded by becoming more reserved and less assertive; as a father, I tended to incorporate my mother's inquisitive and sensitive style of parenting. I requested a certain standard of behavior and order in the household, but my parenting would likely be considered lenient and forgiving rather than commanding. I was drawn to jogging, kayaking, and swimming rather than to competitive sports. I didn't enjoy aggressive male competition.

Relaxing quietly next to Tommy as we watched Frodo, I had time to think about how my relationship with Tommy and my ten-year-old son, Patrick, had evolved. Although Patrick was too young to join us on this trip, he had loved hearing the chimp stories earlier in his life and was confident around animals and in the forests of the Pacific Northwest. We shared a similar tactic in life for coping with stressful situations: escaping into fantasy. For Pat it included dragons and Samurai warriors in books and daydreams, while for me it had always included thoughts of the Gombe forest. We also bonded over using humor to get through tough times.

Tommy was heavily immersed in most of the major team sports from age four onward, and had the ability to stay focused in a soccer or baseball game for long periods. He later branched out to add music (drum, piano, and guitar) to his life, but continued in competitive sports, especially soccer, even in college. My wife, Wendy, commented a few months after Tommy left for college, "I never find the sports section in the bathroom anymore."

I felt sad thinking about that, not only because it symbolized Tommy's leaving home but also because it highlighted our markedly different interests during his formative years. I must admit that the real estate and entertainment sections were my preferred parts of the newspaper. I loved attending Tommy's baseball and soccer games, and even coached his earlier sports teams, but he was a competitive athlete and I wasn't.

On this trip to Gombe, though, Tommy and I were finding common ground. Being dropped into a diverse and primitive setting had brought us closer together. We were comrades on a mission. All the past differences in our interests and childhood experiences seemed to be less important now, as we watched Frodo calmly eating plums in the golden-red sunset light. Tommy seemed completely at peace, as was I. Our father-son relationship began to feel more like being the best of friends.

I could easily see parallels between Frodo and me at that moment: I too was a retiring male (more beta than alpha) and—I hoped—the wise teacher who could share his life experiences with his sons. Though it wasn't clear to me yet when I was a student at Gombe, it still amazes me how much I'm influenced by my days there. Studying early childhood development in college and then observing the chimps in the wild inevitably influenced my image of myself in various roles—from medical student to bachelor, husband, father, and doctor.

The sight of Frodo also sparked memories of old Hugo, a chimp named after Jane's first husband, and how he was treated by the other chimps at Gombe. At age forty-eight, Hugo was obviously no threat to the other males, and was even allowed to scoop a banana out of the mouth of Figan, the alpha male at the time. I never saw him threatened or spurned. Old Flo, the grand-matriarch at Gombe who had died before I arrived in 1973, was also treated with respect. When she was in her late forties, the adult males even found her quite attractive during her estrus.

After about an hour, Frodo rose as his small group stood and started passing by, coming close to us. One by one, with Frodo last in line, they descended rapidly into the thicket to find suitable trees for nesting. Abdul sighed, and my heart sank as I realized the dense vegetation would make it impossible for us to keep up with the agile chimps and see their nest building. Watching them disappear, Tommy's face fell too.

Without saying much, we turned and hiked back toward the beach, watching the purple sky change to a deep blue. As we walked, I started to feel better. Though we never saw nest building, we had met up with Freud and leisurely observed Frodo. I was grateful to see how two of Fifi's offspring had turned out. Taking a deep breath of the fresh air of the forest, I felt deeply connected to the familiar primates. A wave of emotion overwhelmed me thinking about my reunion with Freud, who had helped shape my life four decades earlier. I smiled gratefully at my son's back as he descended the trail in front of me.

CHAPTER SIXTEEN
JANE'S PEAK

After a refreshing night's sleep, Tommy, Abdul, and I met up again in the morning, outside of our concrete building. By eight o'clock, we had eaten fresh papaya and cereal and chatted with Abdul about the day. There was little chance of rain this time of year so when we returned to our room, we dressed lightly and placed only containers of water in our small backpacks.

We set out once again to follow the chimps. Abdul guided us throughout the day as we encountered different groups of chimps, baboons, and colobus monkeys leaping in the treetops, pausing at streams now and then to cool our faces. Toward late afternoon, we decided to visit the peak that had served as Jane's lookout point in her first sighting of the chimps in 1960. I wanted Tommy to see the exact spot, a small clearing overlooking the main valley, where Jane had waited patiently for several months while the chimps gradually lost their fear of her. Several *National Geographic* films show her at this location, observing the chimpanzees in the valleys below with her binoculars and rolling out her sleeping blanket to spend the night.

"I will take you to Dr. Jane's Peak," Abdul said. Abdul knew the entire history of Jane's journey at Gombe, even though it began well before he was born.

By five in the afternoon the forest was calm without a lot of animal calls, and the seventy-degree temperature combined perfectly with the light breeze that blew across the lake. Birds chirped softly all around us when we emerged from the thick brush onto the small grassy bluff called the Peak. Tommy and I were panting hard, though Abdul seemed less fazed by the steep climb.

"I feel like we're flying over the valley in an airplane," Tommy said, looking down on Kasekela Valley. In sharp contrast to the all-engulfing dark green of the forest, the open area where we stood was bathed in golden sunlight. Occasional chimp calls and baboon grunts echoed from the leafy forest below. The three of us stood quietly for several minutes; our silence felt like a moment of prayer or meditation. Standing there was like entering an old cathedral whose stained-glass windows and high ceilings whispered an ancient spiritual message.

"I love this spot in the forest," Abdul confided. I looked at him; peace and serenity shone from his face. I let out a slow, deep breath in response.

For me, the Peak symbolized Jane's hard work, isolation, and achievement during her early years here. Gazing out over the valley down to the lake and up to the Rift Mountains, I imagined her in 1960, at age twenty-six. The loneliness must have been difficult, yet Jane had always emphasized that these were treasured moments. She might have spent grueling days trekking through dense underbrush, risking exposure to tropical diseases and injury, but there were rewards: the peace of the forest and the excitement of seeing the chimps in their natural environment were gifts beyond price and made all the hardship worthwhile.

Jane admitted, however, that she would have been foolish not to worry about the leopards, Cape buffalo, snakes, and malaria, but her self-discipline and determination were powerful. Her strong commitment to the chimps and her work motivated her and allowed her to risk the dangers and endure the isolation far from her comfortable home and family in England.

In her book *Reason for Hope*, Jane wrote, "I was supposed to be there and I had a job to do." She thought she had a pact with God, and she had said, "I'll do this job, God, and You look after me."

Despite huge obstacles and unknown dangers, her dream of working with wild animals in Africa came true. Those genes from her car-racing father and her consistent, caring mother may have been at work. In *My Life with the Chimpanzees* she wrote, "I wanted to learn things that no one else knew, uncover secrets through patient observation."

At a lecture I recently attended, I was fascinated to hear Jane say that looking back at the early years, it had seemed impossible to succeed, but she'd persevered anyway. "I felt strongly after my first few frustrating months at Gombe that I needed to complete the work I was sent here to do." To her credit, after fifty years, her life's work studying the free-living chimpanzees in Kasekela Valley is still in progress. It's the longest-running study of great apes in the wild. Because of its longevity, her research has revealed new information that could not have been discovered otherwise. As a mentor, Jane had a long-term outlook that impressed on me the value of always keeping the end goal in mind.

I was ruminating on all of this when I suddenly thought of something. I had a persistent need to document "special moments," even though I knew that doing so might interfere with my enjoyment of the actual moment.

I handed my video camera to Abdul. "I'd love to record Tommy and me sitting in the spot where Jane sat in 1960," I explained. He nodded, looking down at the device in his hand.

We found the location that we assumed was the exact vantage point for Jane's first observations, and while Abdul recorded us, I began talking with Tommy to create a narrative of the moment. Suddenly there was a loud click. "It stopped," Abdul said, frowning.

Examining it and pressing some buttons, I figured out that the battery was dead. I shook it in frustration. "I thought it had more power," I said. "Maybe I could go back and get another battery." Both my son and Abdul just stared at me in disbelief. I came to my senses; it would have taken three hours round-trip. *Insane.* I was frustrated with myself for not being prepared. While I kept tinkering with the camera, Abdul wandered away a bit and Tommy followed.

The forest has a way of making petty annoyances seem insignificant. One of Gombe's great lessons for me was teaching me to distinguish what was important from what wasn't. It's a lesson I continually learn in my personal and professional lives.

Taking a deep breath, I realized I was focusing on the wrong thing. Instead of being preoccupied with the mechanics of taking pictures and video and narrating "special moments" for people back home, I could simply be *in* my special moments.

I turned toward Tommy and Abdul, who were talking and laughing about their lives and the chimps. I was delighted to see these two young men in their twenties from very different backgrounds joking and enjoying a growing friendship—a scene that could have taken place on any college campus or in the town of Kigoma. It had taken place for me in Gombe, when Hamisi and I started our own lifelong friendship. I was glad it was happening again now before my eyes at Gombe, as it brought me back even closer to that period in my own life. I forgot about the video camera and just observed for a while.

"Back home, some people hire personal trainers to help them get started with exercise routines and weightlifting," Tommy told Abdul as they talked about the still-remarkable Frodo. "Maybe Frodo could be hired as a personal trainer for people who want to be extra-strong."

Abdul burst out laughing and clapped Tommy on the shoulder.

As I watched them compare stories, I felt envious of their youth, but also empathetic about the uncertainties they faced with regard to their careers, finances, romances, priorities, and purpose in life. I remembered my twenties as a tough decade, even with its exciting times. Back then I questioned if I had what it took to become a doctor. I worked hard but found mastering physics and biochemistry difficult. I had good friends in college, but lacked confidence in developing more intimate relationships. Because of my conservative upbringing—and because I was naturally shy and reserved—I found it hard to make easygoing connections with people. Observing uninhibited wild animals in the African forest had been good for me. Some of my own inhibitions and self-conscious feelings had melted away after months of watching the natural, spontaneous interactions of the Gombe chimps. I found it easier to be myself, and I worried less about what people thought of me. In some ways, the chimps' confidence in their world had given me confidence in mine. I learned what it was to be absorbed in an activity that was meaningful for me as an individual while I contributed something important to a larger community. Learning to live life with both purpose and pleasure gave me enormous happiness.

Through his study of Buddhism and meditation, Tommy was currently exploring the idea of living in the moment. He was trying not to dwell on the past or future in his everyday thoughts. It was also clear he didn't want a career track that precluded being "mindful" of others. Abdul too was mindful. He placed great importance on caring for his ailing mother and having a strong faith. In conversation, he was very attentive and present, making a lot of steady eye contact. He and Tommy

ABOVE: In my office in Seattle preparing to see my next patient. *Photo courtesy of John Crocker, 2007.* BELOW: Son Patrick using rebounder to display in mid-air, hoping to land on our bed. *Photo by John Crocker, 2007.*

ABOVE: Reunion with Hamisi on the beach at Gombe after thirty-six years. *Photo by Thomas Crocker, 2009.* BELOW: Tommy and me reflecting at Jane's Peak. *Photo by Abdul Ntandu, 2009.*

Photo Hamisi saved of us from 1973, given to me in 2009 at our reunion. *Photo courtesy of John Crocker, 1973.*

ABOVE: The first chimps Tommy and I suddenly saw on the trail, grooming, on my return trip to Gombe. *Photo by John Crocker, 2009.*

BELOW: Young chimp taking milk apple from mother's mouth. *Photo by John Crocker, 2009.*

ABOVE: Chimp mother with infant, being reassured through touch by adult male. *Photo copyright © The Jane Goodall Institute / by Hugo van Lawick, 1965.* BELOW: Male chimp with water bucket doing a display. *Photo copyright © The Jane Goodall Institute.*

ABOVE: Freud in his golden years at age thirty-nine, relaxing before building his nightly nest in the trees above him. *Photo copyright © The Jane Goodall Institute / by Bill Wallauer, 2009.* BELOW: Tommy and Abdul on the way to Bubongo Village. *Photo by John Crocker, 2009.*

RIGHT: Mother and child, taken in Bubongo Village on our visit with Hamisi. *Photo by John Crocker, 2009.* BELOW: Hamisi with Figan, Tommy, me and extended family in Bubongo Village. *Photo by Abdul Ntandu, 2009.*

ABOVE: A group of Tanzanian women and children on the road to Bubongo Village. *Photo by Thomas Crocker, 2009.* RIGHT: Wind-powered transportation along the lake: hand-crafted sailboat returning from Kigoma. *Photo by John Crocker, 2009.*

were both rugged, fit young guys, vigorous and spontaneous, and both were well integrated into their communities. I wondered what contributions each would make to those communities over his lifetime.

As Tommy and Abdul continued in their lively conversation, I stepped back a bit. Noticing some distant puffy clouds over the hills of the Congo across the lake, I faded off into my own reverie, recalling a time Jane and I had spent together a few years earlier.

Jane had come to speak in Seattle, and in midafternoon, she and I went out onto a terrace for a glass of wine. It was a mild day, and we enjoyed sitting in silence for a while, enjoying our own thoughts together. I watched a squirrel rustling around in the cluster of trees in front of us.

I was watching the branches on one of the trees, which were moving gently in the breeze when Jane broke the silence. "I have a tradition of making a toast to the people in my life who have passed on" she told me. "Each day at around five P.M., I think about my mum and about Derek." Jane's second husband, Derek, had died from colon cancer only five years after they married. Her "mum," Vanne, who had nurtured her in childhood and provided emotional support during her early days at Gombe, had died a few years prior, at age ninety-three. They had been very close.

"I have lost so many good people through the years," Jane said. We were quiet again for a while, and then Jane started talking about other loved ones who had passed away. She spoke of them with warmth and with joy. She talked about Hugo, her first husband, who had also passed away too young; Danny, her grandmother, with whom she had spent much time; and of course Flo, the chimpanzee matriarch, whom she had come to know well in her first years at Gombe.

I looked at Jane, whose face was still and serene, and moved over closer to her. Remembering her effectiveness with nonverbal communications, I simply gazed at her and smiled. She did not need me to reply.

Jane contemplated the deep-blue sky with puffy white clouds and said, "I guess it's time for a toast to the Cloud Contingent."

I came to appreciate her inner spirit even more as she described her way of keeping in touch with those who were no longer part of her physical life. Each day since their deaths, Jane told me, she thought about these precious people in her life, which she called her Cloud Contingent. The cloud formations were like her connections with these loved ones, constantly changing but always there. Her closeness to them and to nature seemed tightly intertwined.

So as the puffy clouds blew by, Jane and I toasted: "To loved ones who have passed."

Now, recalling that afternoon, I watched from the Peak as the clouds moved across the hills of the Congo. Reaching out, I grabbed a large leaf and made a primitive cup out of it. Tommy and Abdul, startled by my sudden movement, looked at me as though I'd been out in the sun too long, but I held the "cup" pointed up to the clouds, saying, "Here is to life on Earth now, to Jane for all her hard environmental work, and to the Gombe chimps."

Tommy and Abdul both flashed me big smiles, and they held up their hands as if holding cups to toast with me. My son looked at me with appreciation in his eyes as he held his imaginary cup to his lips, and my heart was full.

Looking around us from our spot on Jane's Peak, feeling uplifted and surrounded by divinity, I thought about my mother. I had first learned about spirituality from her. While I was growing up, she read books about Edgar Cayce, a spiritual healer who believed in reincarnation and predicting the future. I remember being skeptical but also interested in the stories she told me about Cayce discovering answers to medical problems in patients while he was in a trancelike state. Grounded in her faith, my mom had a calm and confident response to emotional turmoil and health concerns in our family. "Just turn within and find the truth," she would say in a trusting voice. "God is within each of us."

Later I learned more about spirituality from Jane, who shared my belief that the forest was the most divine of places. While at Gombe, I could see the depth of Jane's spiritual life in her observations and inter-actions with nature and wildlife. I felt an echo of that in my own soul then, and I felt it now, on the Peak. As Dale Peterson wrote of Jane in *Africa in My Blood*, "As she came closer to nature and animals, she came closer to herself and in tune with the spiritual power she felt all around."

After I left Gombe, Jane had told me of a difficult time in her life when, in 1974, her marriage to Hugo was breaking up and her affections were moving toward Derek. One day in the spring of that year, while feeling despair and sadness, she happened to walk into Notre Dame Cathedral in Paris. She was immediately surrounded by the thunderous notes of Bach's "Toccata and Fugue in D Minor," playing at full volume for a wedding. The music throbbed around her while the sun illuminated the great Rose Window, and Jane paused "in silent awe." Peterson, in *Jane Goodall: The Woman Who Defined Man*, beautifully described the moment, and Phillip Berman, in *Reason for Hope*, also captured Jane's thoughts in the cathedral:

> It was the glorious reverberation of the organ in an ancient place of worship, sanctified over hundreds of years by the sincere prayers of so many thousands of people. The impact was so powerful, I suppose, because it came at a time when so much was changing in my life, when I was vulnerable. The experience forced me to rethink the meaning of life on Earth.

Standing there on the Peak, I remembered those words. I too was standing in a place that had an ancient spiritual imprint. Looking out on a forest that was millions of years old had a profound spiritual charge to it. Getting to intimately know the lives of the animals that lived there and having had the privilege of seeing our human connection to our chimp cousins there gave me a sense of both the vastness of the world and of the specifics of our moment in it.

My feet felt rooted to the rich soil; I stood there with my son, who was seeking his own way into a more spiritual and connected life. I felt profoundly grateful for both Jane's and my mother's earthly and spiritual influences, and I was filled with the wonderful awareness that I was seeing my son experience this remote landscape and history in his own manner. It was natural at that moment for my thoughts to turn to another female role model, no longer alive on this earth, who had influenced and inspired me as well—the magnificent matriarch Fifi.

Remembering Fifi

While Tommy and Abdul were contentedly talking and starting to explore the Peak together, I wanted to take a little more time to reminisce on this historic spot. I called to Tommy, and as he approached I said, "I need a few minutes to think about Fifi."

I meandered over to a knoll of dry grass near a fig tree, where I could sit and reflect upon Fifi and her effect on my life. How lucky I was to have been able to come alongside such a highly successful chimp mother raising her first offspring. Lucky, too, that my observations had taken place during eight months of Freud's formative years, when a lot of learning occurs. Fifi's remarkable mothering skills—and how we humans might learn from them—were often in my mind when I began to work with patients.

I couldn't help but have some guilt that I hadn't seen her again. For many years before my return trip with Tommy, I kept telling myself, "I'd better go back before Fifi dies." At age forty, chimps start to show signs of aging and are more susceptible to pneumonia and other infections. Some may live to fifty in the wild, but not many. I really wanted to see Fifi again.

In July 2004, however, before I could return to Gombe, Tony emailed me at work, telling me that Fifi was missing. Doing the calculations, I realized she would have been close to forty-six. I knew Fifi well enough to know that she would not be missing for anything. She must have died. Fifi, the successful mother of nine, daughter of old Flo, focal point of my student observations because of her superb mothering skills, energy, and confidence, was no longer alive.

I felt terrible. I told myself, *How can you feel guilty about not seeing Fifi before she died when you meant nothing to her?* Since researchers didn't interact with the chimps (unless it was unavoidable), I had probably been just another strange, upright ape to her. But for me, the five stages one goes through after the loss of a loved one held true even in a one-way relationship with a wild chimp: disbelief or shock, guilt, anger, grief, and eventual acceptance. Our sense of loss isn't always mitigated by how much we meant to someone we lost—it's about how much they meant to us.

One month after hearing the news, I wrote Fifi a long letter about how proud I was of her accomplishments and how much she'd taught me about mothering. I wanted her to know how much she had influenced me and how I continued to pass that influence on to patients who have young children.

Sitting on the Peak, I remembered that evening in the office, looking up at her picture while getting out the pad and paper. I thought she would prefer it handwritten.

Dear Fifi,
Although physically not present on Earth, you are still often in my thoughts. You lived forty-six amazing years in the wilds of the Gombe forest.

For eight months I watched you as a young mother raising your two-and-a-half-year-old son Freud. You were my favorite of the four mothers I studied because of your endless patience, playful manner, and overall confidence in life.

To the astonishment of everyone at Gombe, including Jane, you successfully raised nine offspring over a thirty-year period: Freud, Frodo, Fanni, Flossi, Faustino, Ferdinand, Fred, Flirt, and Furaha. Three of your five sons achieved alpha male status and helped maintain the community's protection and vitality for many years.

You were an engaging chimpanzee like your mother, Flo, and won the hearts of people around the world who saw you in *National Geographic* films and read about you in Dr. Goodall's books. Your notoriety has contributed to more awareness of the urgency to conserve forests to protect threatened species such as chimps, orangutans, gorillas, and other precious animals.

To you, I was probably as significant as another bush you passed while journeying through the valleys at Gombe. A variety of curious human primates had been watching you since your infancy; however, you brought tears to my eyes when I pictured you struggling to nurture two-year-old Furaha during your final days of life. I would bet that you expended as much energy trying to keep her alive as you did fighting for your own survival. Perhaps Freud or Frodo was nearby to help you.

You comforted me while I was at Gombe by showing up at my hut with Freud on Christmas morning in 1973. What a delight to walk out the door and see you two playing in the sun ten feet from my hut. Better than Santa Claus! Your exuberant play sessions with Freud reminded me later in life to spend lots of time wrestling, hugging, and just hanging out with my two boys. I tried to be patient with them and understand their periodic wild behavior. Our house has seemed more like a jungle at times than a civilized dwelling for bipedal primates.

When I heard of your disappearance last month I took a momentary time-out from being a busy doctor and father and husband. I had really wanted to see you before you died to reconnect and observe the changes in both of us over the three decades since I last saw you. No one knows how you died or who was with you at the time.

You are a legend to many. You are a hero to me. I will always remember how vigorous and competent you appeared the first time I saw you stride into the upper camp with Freud clinging to you tightly. I could tell that you were going to be around for decades to come.

I hope your offspring and grand-offspring will thrive as you did in the forests of Gombe. I'll never forget the privilege of studying you in the wild. I doubt most people would envision learning life skills from a fifteen-year-old chimpanzee mother living in the wild, but I am living proof.

From the heart,
Bwana John

It may seem silly, but writing the letter helped ease my sadness. I had needed to articulate my feelings, and the letter seemed a good way to accomplish this. And now, another way of easing my grief was to spend time on this return trip observing two of Fifi's sons: Freud, thirty-nine, and Frodo, thirty-four. Frodo's successful hunt on this trip showed me that despite his age, he still displayed vestiges of his powerful alpha male days, reminding me of his mother's incredible stamina while raising her nine offspring.

Freud was the only chimp I was able to see in 2009 that I had also studied back in 1973. When I was reintroduced to him, one of the newer field assistants smiled at me and asked, "Do you still recognize him after thirty-six years?"

I looked intently at his face, as he sat there, eating. I flashed back to him as a youth, swinging from vines and playing with Gremlin, and I suddenly saw it all: postures, expressions, facial features, and subtle mannerisms—they all help distinguish the individual chimps, just as they do humans. "Yes," I said happily. If I observed closely, I could likely have picked him out of any chimp lineup.

As I watched Frodo and Freud on this trip, I had seen in these two confident chimps the competence and wisdom of their mother. Although we don't know the proportion of genetic versus environmental influence on upbringing, I am certain that Fifi's exceptional mothering was vital in enabling her offspring to succeed.

Now, sitting where Jane sat nearly half a century earlier waiting for the chimps to accept her, I looked out over the valleys one last time. I felt complete (*nimeshiba* satisfied) and ready for Tommy, Abdul, and me to leave the Peak. As we began our descent into the forest, I felt calm yet energized after reflecting on Fifi's life and what I'd learned from her. While we hiked and enjoyed the views of Kasekela Valley, I started telling Tommy a story.

"On one of my days off, I spotted Fifi near a hut just up from the beach," I explained. "The chimps usually stayed high up in the valley, but there she was! I saw her reach through an opening in an otherwise screened window of the hut and snatch a plaid shirt belonging to one of the students." Tommy laughed and Abdul chuckled quietly.

"First Fifi draped the shirt over her back. After she walked around like that for a while, she used it to play tug-of-war with Freud." I couldn't help but laugh at the memory. "Then she chewed on it and dragged it around until she finally discarded it on her way back to the valley."

Tommy said, "Oh man! Good thing it wasn't your shirt!"

"I wish it had been my shirt," I owned. "Then I could have shown it to you kids when you were little and told you the story of a legendary chimp named Fifi slobbering on it during a tug-of-war game with her son."

"Good old Fifi," my son said, smiling.

CHAPTER SEVENTEEN

FINDING MY PLACE

Through the course of the day, Tommy, Abdul, and I had traveled across several valleys, observed three different groups of chimps, and spent an hour at the Peak. At 6:00 P.M., however, we couldn't find a single chimp to follow for nest building. As the sun slipped toward the horizon, Abdul and I knew we had to work fast to catch up with any group of chimps if we were to succeed in our quest to watch them settle into their nests for the night.

I really wanted Tommy to witness the nest building. The chimps demonstrate extraordinary concentration and experimentation when they perform this essential survival task, and it reveals their amazing brainpower. Weaving the growing branches of a tree into a hollowed, springy platform high in a tree is an impressive feat. Occasionally a chimp remakes an old nest that happens to be there, but either way, most of the branches making up the nest remain growing, and the forest shows little sign of alteration from the activity.

Abdul led us quickly through the underbrush, and I told Tommy, "Keep your eyes peeled." I was feeling anxious—what if we missed this important ritual again?

Soon, just below the Peak, pant-hoots rang through the forest. We stopped and looked around, but couldn't see the chimps. "*Twende!*" Abdul commanded. "Let's go!"

We began to scramble across the valley floor at a running pace. The toughest portion of our mad race through the jungle was scampering up a steep hillside while ducking low-lying branches. Forget the mambas; we had to just keep our eyes straight ahead to avoid getting scraped by bushes or trees. Tommy laughed and I looked up at his retreating back; he seemed to be relishing the intensity of the moment.

After forty-five minutes, Abdul and Tommy were neck and neck at the finish line, while I was fifty yards behind and hoping not to get lost. When we finally reached the top of the hill, I was breathing so hard I couldn't talk. Tommy gave me a huge smile, and Abdul signaled a thumbs-up when we found ourselves completely surrounded by fifteen chimpanzees, on the ground and in the trees, some only ten feet away. I would have given a sigh of relief if I had been able to catch my breath.

"Stay put," Abdul whispered as we looked at him, wondering what to do next. At that moment, a young male researcher, moving quietly and carefully trying to blend into the landscape, also arrived on the scene and sat down in the grass to observe.

I looked up and saw twenty-year-old Pax on branches right above me. Then Frodo strutted by, almost touching me. I felt a thrill travel down my back. Freud sat fifteen feet away, leaning against a tree trunk. We had entered their gathering without realizing it, and they seemed completely disinterested in us. The pant-hoots continued as the group began to shift and move, looking for suitable trees for nest building.

"What do you think?" I asked Tommy, who appeared spellbound.

"Amazing," he whispered back, not taking his eyes off Frodo.

Another adult male came within five feet of me and began to scale a tall tree. He reached around it with his hands and walked up it with his feet.

"Is he building a nest?" Tommy asked.

Abdul said, "He could be, or perhaps he is looking to see if other chimps are nearby."

We slowly tagged along behind Freud, who had begun to move away from the crowd. We eventually stopped halfway up a hill to watch a nineteen-year-old male named Zeus build his elaborate bed for the evening. He was twenty feet above the ground, but we were able to observe him at eye level because of our position higher up on the hillside. It was an extraordinary view. We quietly watched Zeus weave branches together with his hands, then mash them down with his feet. Periodically he would lean back and survey the growing nest and make appropriate additions and adjustments.

From one of Tommy's journal entries that evening, I was able to get a feel for his reaction to the scene we witnessed:

> The sun is low but still glowing orange through the trees. A gentle, balmy breeze sweeps off Lake Tanganyika, ushering in an undeniable calm. A group of about fifteen chimps goes about their evening routine, relaxing, scratching, and eating fruit high up in the trees. My dad, our guide, a researcher, and I stand in the center of this dusk ceremony, paralyzed by awe, afraid to move. One chimp ambles up the path headed directly toward me. He brushes my leg as he passes—an accident or a greeting?—before completing his journey to a resting place nearby. Some chimps groom one another by picking through their companion's hair, while others simply rest.

As the light fades, a slow procession begins. One by one, the chimps select a tree, climb up about twenty feet, and bend the limbs—a nightly ritual. While most chimps work outside our line of sight, one, Zeus, gives us a glimpse directly into his creation in a low-lying tree. He is deliberate, taking care with each branch he bends.

After about ten minutes, he apparently finds the nest suitable and rolls onto his back to stretch out. As the others finish their work, Zeus emits high but gentle hoots, different from the anxious yells and hollers we heard throughout the day. Others respond softly and the air is filled with this mild banter.

Eventually all is quiet. We have to leave to get back to camp before the darkness takes over. As I leave, I hope to take with me the sense of peace and ease that I saw in this community of chimps.

After they were settled, Tommy and I left Zeus and the other chimps for the night and walked down to the beach with Abdul. We could see the fishing boats far to the north and feel the mild evening breeze. This had always been my favorite time of day at Gombe. Tommy and Abdul explored a bush with some ripening fruit and I stared out at the lake. I remembered being Tommy's age, swimming, taking the boat into town, and enjoying the sunsets in the evening after a hard day's work.

This moment could have made me melancholy. So much had changed. I was no longer a naive, spontaneous young man, Jane wasn't here on the beach with Grub, and so many of the field-workers I knew had retired. But instead of thinking of the past, I decided to take a deep look into the here and now. I watched Tommy and Abdul comb through a bush and taste the ripening fruit they found, listened to the calling of small birds near the lake, and watched as the sky changed color with the setting sun.

I was part of the larger landscape, as I had been during my final months as a student at Gombe. I had no desire to try to control or direct the world around me. Instead, I saw myself as a privileged guest in an animal kingdom that tolerated my presence. There, on the beach, I was an observer of not only the chimps' world but also my son's and my own.

I realized that one of the reasons I felt so at peace at that moment was that the chimps and the forest had helped me understand my life more completely. The chimps' behavior had somehow helped me begin putting together the missing pieces of my own childhood and growing up with an alpha male father. My observation of these primate cousins added perspective and contrast to my life.

My dad had worked hard as an executive in two different companies. I observed his drive and the stress that went with it, similar to what Figan, Freud, and Frodo endured to become number one in their community. Though my own more reserved nature didn't lend itself to alpha male challenges, I recognized that a portion of my makeup included grandiose ideals from an early age. I saw myself contributing to society in a major way, but less conspicuously. During high school, I was a hard worker. Doing well at school was important to me, but I wasn't interested in sports, the school newspaper, or even band, so my dad urged me to run for a leadership position instead. The only position none of my friends was running for was student body president, so I reluctantly ran. Even this was a level of competitive engagement that was out of my comfort zone. I didn't have a real platform, but I did create very artistic posters. To my surprise, I won.

Perhaps I was like the resourceful Mike, who, rather than fighting, became alpha male by banging kerosene cans to intimidate others. After all, my artistic and humorous campaign posters may have pushed me past my competition. I had run against the debate captain, who could speak forcefully and with great determination. My gentler approach may have created a sense of trust among my fellow classmates. Rather than conform to the standard model of a student body president, I had presented other traits that my peers responded to positively. It was a good lesson—to learn that I could succeed in a competitive environment without being a competitive person by nature. I had other resources that I could rely on. I enjoyed the leadership position and continued to incorporate creativity and inclusiveness while serving.

Watching the chimps in their communities helped me recognize that my drive to accomplish something significant might be related to those genes that I saw in both my father and Figan. Even the chimps—male and female—that don't reach a high rank in the community may still have a strong drive to achieve other goals. Faben, for instance, seemed driven to support his younger brother in reaching the top rank and in helping him hunt successfully. He sought a supporting role.

My mother was a tremendous influence and helped model how to be a good parent. She wasn't a competitive person by nature either. She was highly sensitive and always anticipated our needs. She instilled a feeling of comfort by giving us reassurance in times of illness or trauma. At age ten, after I fell onto a curb while roller-skating and severely broke my arm, she calmly summoned our neighbor to help get me to the hospital. She was at my side the entire time in the emergency room reassuring me and explaining what was happening, never getting anxious herself. Her nurturing and confident style made everyone in the family feel secure. That sense of security has been a lasting gift. My mom was attuned to our physical and emotional health, and she mirrored some of Fifi's best mothering skills—attentiveness, patience, and encouragement.

As a father, I wanted to instill that kind of confidence in Tommy and Patrick. Being a physician, I was always alert to the smallest physical changes in them—perhaps overly so. When they were young, I would worry that the slightest fever indicated a disastrous infection. I tried to keep that fear—and other doubts—to myself as much as possible. As any parent knows, the world can seem fraught with danger at every turn. Every sharp table edge, every hot cup portends injury.

In the forest, though, on both of my trips, I didn't obsess or worry about myself as much as I had thought I might. As the natural rhythms of the plants and animals become your mirror, you feel more certain of yourself. Organisms display their amazing capacities to heal and adapt to injury, adversity, and illness. Simply put, when it rains, you know you'll get wet, and you also learn that you'll eventually dry off. This time I was happy to also see Tommy confidently adapt to the multitude of new experiences he encountered: a new environment, new people, new species, and a new language.

Standing beside the ancient lake at dusk, I felt profoundly content. Watching Tommy and Abdul as we shared the forest with our closest relative had added to my sense of personal peace. Understanding my own place in the big evolutionary picture was reassuring. I found it comforting to think of my son and myself as tiny links in a long, long, evolutionary story. It made me feel more closely connected to the chimps than ever. The spirit of the forest was alive within me again, and I knew that I would take its wisdom home and integrate it into my life and profession in deeper ways.

I realized during my forest walk that Tommy and I both felt a stronger connection to nature and a sense of a more primal world with each passing day. As I had done on my first trip, I pondered how to incorporate some of this primal nature into my life back in Seattle. I joked with Tommy about planting grass or giant ferns in my exam rooms along with some miniature palm trees. In fact, my officemate and I do have significantly more plants in our office than any of our clinic colleagues. But my Gombe world and my Seattle world seemed far apart at that moment. I knew when I went home I could escape to the Cascade Mountains or daydream about the African forest, but now I wondered what new insights still lay ahead—what would I take home from my current forest journey?

CHAPTER EIGHTEEN

FOREST REFLECTIONS: HOW WILD CHIMPANZEES HELPED MAKE ME A BETTER FATHER

We found the slab of concrete in the forest, and my heart sank. I wanted to question it, but I knew that this was it—I remembered the precise location of the place I had lived for eight months. A part of me had hoped it would still be standing. "Well," I said solemnly, "this was my hut." Abdul, Tommy, and I stood there and stared down at the dirty, twelve-by-ten-foot concrete rectangle. It reminded me of an abandoned dwelling in the countryside with grasses and bushes growing all around it, but no walls left. Or perhaps ancient stone ruins.

It was our fourth day at Gombe. I had rejoined Tommy and Abdul after a solo forest walk and told Abdul I really wanted to see the hut I had lived in as a student, so he led us up the path to the chimp camp, a fifteen-minute hike into the forest. On the way, Abdul told me, "You should know there are no longer huts in the forest for researchers to live in. They decided everyone should live down on the beach, away from the chimps." This change was meant to reduce the impact on chimp behavior. And it was for staff safety as well, following the May 1975 kidnappings of students whose huts had been more isolated in the forest. The research center was keeping to national park guidelines and reflecting local cultural wisdom, namely that safety is found in living close to your neighbors. Still, I felt a twinge of sadness when I realized that my small thatched aluminum structure with a view of the lake was gone.

I looked sorrowfully at the hut's foundation. I had really wanted to go into the hut, sit at the desk, and maybe even write a letter home—and I had wanted Tommy to see the hut. Walking over and stepping up onto the concrete, I gazed out through vines and low-hanging branches to catch a glimpse of Lake Tanganyika, remembering this exact view from the hut's window.

Abdul joined me on the slab. "Where did you sleep?" he asked, and I walked him over to where my bed had stood, pointing out where the door and the window had been as well.

Crossing the slab again to where the window had been, I spread out my arms, telling Abdul and Tommy, "My desk was right here. I would sit here at the end of each day and fill out my reports and then write letters to my parents." Describing the candles I had used for reading and writing, I realized I was growing excited again from recapturing these moments.

While Tommy walked around the concrete rectangle, peering through the imaginary door and window, Abdul said, "It is interesting how basic your accommodations were for those eight months." Reflecting back to the modern researchers' housing on the beach, including their flat-screen television, I nodded and chuckled.

Even though the structure itself was gone, I liked knowing that Tommy could now more clearly picture my life back then, living in this remote area, far from sight of most civilization, and alone from the time I entered my hut each night until the next morning.

Standing with Tommy where my hut once stood, I was overcome by a strong sense of reconnection to Tommy's childhood. I looked at my son's young and curious face and recalled how, when he was a child, I would tell him bedtime stories about my Gombe adventures.

That evening Tommy wrote down his thoughts after visiting the site:

I felt ambivalent as I re-created the hut in my mind from pictures I had seen and stories I had heard. I had first romanticized the individualistic ascetic lifestyle. Harkening back to the first fifty pages of Thoreau's *Walden*, which was all of the free reading I could manage during my sophomore fall semester at college, I couldn't help but envision a spiritual and moral strengthening that surely would arise from the absence of consumerist clutter. Wouldn't experiencing those things most fundamentally human—survival amid danger, finding food and water—as well as the vivid immediacy of weather, lunar cycles, and death and decay, bring out some kind of primal power that seems to be lost in modern civilization?

But then I remembered the story my dad told me about a green mamba hanging down from the ceiling of his hut when he returned one evening. And beyond the threat of poison and sharp fangs, wouldn't there also be a crushing loneliness? I recoiled at the thought of such minimal human contact and living without neighbors and friends a short walk or phone call away. Going from a college environment where students are packed together in dorm rooms like factory farm animals to an isolated Tanzanian jungle mountainside seemed utterly frightening. However, my dad, who had experienced that specific transition, exhibited only joy upon returning to the site of his hut, so I couldn't help but slightly favor the romanticized version of life in the wild.

As we all hiked along in the forest afterward, I realized that in some ways my current hut, my home in Seattle, has included features inspired by Gombe. As our children entered our lives, I automatically geared the house for active young primates. My wife, Wendy, added her more refined interior designs, and the boys eventually added their own creative touches, such as fluorescent galaxies glued to their bedroom ceiling.

Our backyard includes a portion of an abutting wooded area, where we constructed two tree houses and placed three rope swings. After watching the young Freud show off his skills high above the ground at Gombe, I wanted my own kids to be equipped for arboreal life.

Inside the house, our bedroom has a high ceiling, so it became a basketball court—complete with portable net by day and a "nesting place" for my wife and me at night. Our family room served as a gymnastics-dance center, and we share it with our dog, fish, and birds. With my wife's music—singing, keyboard, and guitar—and Tommy's drum set, and a big driveway for skateboarding and basketball, the house became a teenage gathering spot with ever-changing activities. By age seven, Patrick had introduced break-dancing into our home along with guinea pigs, archery, and Lego-mania. Since he is nearly eleven years younger than Tommy, he took over tasks his big brother had outgrown, such as serving as "chief" of the rope swings when neighborhood kids showed up in our yard.

Though spontaneity and creative chaos ruled most of the time, we had peaceful dinners and usually had time to relax together in the evening. The basketball marks on the bedroom wall and dents in the floor molding are pleasant reminders of those early years with the boys. I think I did Fifi proud with the way I chose to overlook the odd mess or disruption to the physical environment.

My fathering style incorporated more of my mother's and Fifi's influences, but I did fall back on my father's authoritative discipline in difficult circumstances. When Patrick needed a time-out at age three, I remember being totally incapable of using diplomacy or gentle cajolery with him as he was so wound up, so I picked him up, his piercing vocalizations rattling my eardrums, and placed him firmly in time-out. I hated the feeling of overpowering him, but I envisioned myself as Figan displaying his firm leadership as I carried out the task. I understood at that moment that as the adult and alpha, I had to exert both authority and leadership—it was part of the order of life. Patrick seemed to understand this, and luckily I didn't have to do it again.

Perhaps in reaction to my highly disciplined upbringing—and partly because of Fifi's nurturing influence—I tended to be lenient with my boys. I used my understanding of our primate nature to justify their occasional uncivilized eating habits and raucous wrestling in various rooms in the house. We always bought sturdy furniture and reminded our boys to restrain their wild side when visiting their friends' homes. Loud vocalizations would ring through our house and the neighborhood as they swung from our rope swings and played guitar and drums in our basement. Figan had used all fours to drum on tree buttresses to create his symphony in the African forest; if you count the foot pedals on the drum set, Tommy kept up with Figan's quadrupedal power. It seemed as natural to Tommy as it had to Figan.

Looking back, it might have been beneficial to instill a bit more order in the household and create more calm moments, but at the time I felt that my most important role was to be available to my sons for emotional support as they traversed the stages of development. My role wasn't to teach them how to use a computer or speak Spanish, since by age seven they were ahead of me on both counts. Wendy played other key nurturing roles, and introduced the boys to music and literature. She always made herself available as a listener when they needed to talk.

Both of us wanted the boys to take more responsibility for cleaning up their messy rooms during adolescence, but we attended a lecture by a specialist in teen behavior, and she said, "Pick your battles," reminding us, "Having your kids keep their rooms clean may not be the most important battle." So we just agreed to ask them to keep their doors closed during those predictably unpredictable teen years. Wendy made surveillance missions once a month to make sure nothing alarming was growing where it shouldn't be. And we let them know that when they became adults and got roommates, they would need to be more civilized and orderly.

Another realization I had as a father, physician, and chimp observer was the importance of "letting it out." Figan's and Satan's mighty displays caused the ground to tremble beneath my feet. When they thundered by, I could tell they were releasing stored-up adrenaline in dramatic behaviors. At the waterfall at Gombe, the loud crashing of the water as it hit the rocks and stream below seemed to stimulate aggressive displays. During one particular rainstorm in 1974, the chimps swung on branches, threw palm fronds, and looked as if they were performing a prehistoric rain dance. Now, on my return visit, Tommy and Abdul were spontaneously inspired to perform some of these same displays, perhaps because they knew the male chimps behaved this way after arriving at the waterfall. It looked so natural and typical of a human reaction that I again thought of the similarities in our species and laughed.

Also, I came to appreciate the other side of the spectrum in my boys: their caring, emotional spirits. As we headed back to camp after visiting the site where my hut once stood, Tommy said, "Remember the nest story you'd always tell me and Patrick?" I had told the boys a true tale based on the night I actually spent in a chimp's nest during my student days:

As the sun began to set, I knew I had to quickly find a nest high up in a tree before dark, just as the chimps do. Luckily I found one that was not too far out from the tree trunk, and I managed to carefully climb up the tree and slowly crawl across some strong branches to get into the nest before dark. The chimps were far away, but I could hear them calling to one another as they said goodnight and settled into their nests.

Unfortunately, my nest was built a week earlier by ten-year-old Goblin, so it was small for me and not very comfortable. It was also lonely being up some forty feet in the tree all night far away from other people and the chimps.

One night, Patrick, at age five, had added to the story:

Then a baby chimp who was lost from his mother thought you were another chimp and crawled in with you. The baby chimp kept you company, and you kept the baby chimp warm and protected, cuddled up next to you.

That fictional addition by Patrick was from the heart. Looking at his earnest little face, it struck me with full force that the scariest element for him was not the possibility of falling out of the tree but instead my being alone, so he imagined having another frightened and lonely primate find me. We would both be consoled.

From then on, after their addition to the story, I would always continue the story, transitioning into full fiction but including true emotion:

The next day the baby, named Prof, jumped up and down on my tummy to wake me up. He wanted me to help him find his mother, Passion. I climbed down the tree with Prof clinging to my ribs just like Babu used to do.

As we searched the forest for Passion, the strong chimp Figan heard us and performed a powerful display that frightened Prof so much he almost broke my ribs from clutching me so tightly. We ducked under some bushes and hid until Figan left, and then continued our search for Passion.

We finally found Passion, and even though she was known for not being as caring as the other mothers, she rushed up to Prof and hugged him tightly. As they went deeper into the forest together, I returned to the camp to find my fellow students and field assistant.

The real story was that for most of that night in the nest, I was frightened, uncomfortable, cold, and restless, and I felt no sense of a romantic adventure except during the first half hour, as I watched the sun set over Lake Tanganyika. After that I was mostly miserable. It took two days to recover from sleep deprivation and a sore neck. I did not repeat the experience.

I had hoped to feel a closer connection to the chimps by remaining out in their territory all night, completing a twenty-four-hour cycle from sunrise to sunrise. I wanted to sway in my bed suspended high above the ground and listen to the wind blow the branches. I wanted to see the stars and watch the vanishing lake as the brilliant sky turned to darkness. I wanted to feel the mystery of the forest with the occasional grunts of bush pigs below and the never-ending symphony of crickets. I had felt a bit sad every evening when I left the chimps after they nested. I would always make my way back to the beach camp to join the other humans, but I missed the chimps. I wanted to be part of their community for just one night, and wake up with the morning sun and the sounds of birds and chimp calls.

My heart still remembers the wonder of that nest adventure, just as my mind retains the cold reality. I included both heartfelt memories and challenging realities in the story I told my sons, and they added more heart and even tried to comfort me.

The common thread in the bedtime stories, in my memories, and in my relationships with Babu, Fifi, my boys, and my wife was the heart. I found spirituality in the forest at Gombe while I found emotional satisfaction in the bonds I formed with the people and chimps there and at home. These bonds continue to influence my life.

So as I followed Abdul and Tommy down the path, I made another toast to Jane for believing in herself when she continued to name the chimps early on in her work, and not number them as many people in the scientific community told her to do. It was important in conveying the full measure of these primate lives to the public. It made the chimps real to the world—it made them live in people's hearts as well as in their intellects. For me, it would have ruined those bedtime stories if I could only talk about chimps #37 and #42 interacting on their daily travels through the forest.

I imagine most parents picture themselves acting like their parents. Parents are our main role models. However, in the complex and sometimes frustrating journey of parenting, I discovered that it can be important, especially for teenagers and young adults, to have exposure to role models outside the immediate family. My grandfather was a key male role model for me early on. In my early twenties, I was fortunate to have Figan, Fifi, and Jane to observe as they raised their offspring. Watching them, I learned what it meant to let go a bit, and to understand wilder parts of our human nature that I might have otherwise missed. I learned to let myself be absorbed into family life in a way that was organic and open. Our house might still have been creative and lively without their example, but probably not quite as much fun.

CHAPTER NINETEEN

A PATH LESS TRAVELED: BUBONGO VILLAGE

Tommy and I were quiet as we walked through Kasekela Valley on our way to Hamisi's house in Bubongo Village. It was our fifth day at Gombe. The trail we took seemed to reflect life's journey. It was winding, with steep ups and downs, sometimes secure and sometimes fragile, with loose gravel on the mountainside, leading to unexpected places and sometimes to unexpected people along the way. The path was dry and faded brown in some areas, but rich and orange-red in the fertile valley near the village.

Tommy and I were accompanied by Abdul and Rudo, a nineteen-year-old fisherman selected by the park director to help guide us. We four would ascend to the five-thousand-foot summit of the Rift Mountains, descend into the fertile valley outside the park heading east, and then trek up gentle hills to the village where Hamisi grew up and now lived with his wives, children, and grandchildren. I hoped I might see other field assistants I'd worked with during my student days. We would return later the same day; the entire journey was comparable to a mountainous half-marathon.

What struck me was how out of place Tommy and I were. We were trekking to a remote African village along a trail used only by locals, with two Tanzanians whose company we enjoyed but had just met a few days before.

Having Tommy with me to share the experience was gratifying since I knew he would appreciate a "road less traveled" at this time in his young adulthood. Many of his friends back home had roots in different countries, and he had enjoyed his own cross-cultural experiences when we traveled and went on work trips as a family. Maybe we were more "in place" here than I had thought.

"I've thought a lot about this visit to Hamisi's," I told Tommy as we walked. "I have no idea what people in the village will think of our visiting them. I'm sure Hamisi will appreciate it, but I wonder if some of his family will be uncomfortable with two strange white dudes and two unfamiliar Tanzanians showing up on their doorstep." During my first visit there, it had been just Hamisi and me visiting his parents.

Tommy replied, "I'm sure Hamisi will make it all work."

This would turn out to be an understatement.

For now, though I felt self-conscious as a mature Western doctor marching along an African forest trail with three guys each a third of my age, I cherished the chance to connect with Abdul and Rudo in their land and to have my son experience village life. Also I felt less inhibited in the jungle, where I could free myself from the self-restraint I'd learned earlier in life. There was no need to please or be a peacemaker; I was able to be more spontaneous and was less concerned about what other people thought of me. The warm, humid climate, the animal calls in the distance, and the Swahili spoken by Abdul and Rudo allowed me to break out of the constraints of my culture and surroundings back home. I felt naturally free to be myself.

As we approached the summit, I found myself singing "Over the Rainbow," which always brings me joy. In a way, the song was quite fitting as I looked out over the tops of crooked-armed trees down to majestic Lake Tanganyika. After some deep breathing at the summit and taking in the views, we headed down into the valley on the eastern side of the mountains. I found myself smiling a lot and laughing more than usual at the wisecracks my adventurous companions made.

I felt anxious as I anticipated entering Hamisi's world. We passed by streams, small corn and cassava farms, and people tending their crops. A boy about seven years old herded goats on a hillside. I couldn't imagine any child his age in the United States with such responsibility, yet this confident lad with his tall staff and a determined look on his face was definitely in control of his animals. He kept his eyes on his work as we passed by on the path below him.

Images of Tommy at that age in his various sports jerseys flashed in my mind. He had played basketball, baseball, and soccer every year from age four to fourteen, and then soccer year-round until he left for college. Was I the dad I'd really wanted to be back then? I thought about how much time my medical practice had consumed and wished I had planned for more family time. We went on family camping trips and other adventures together, but our lives were too busy to spend more time out in the wild.

Thinking about the young goatherd we had just passed, I wondered if my wife and I had been overly protective of Tommy; maybe we could have given him more responsibilities. Is that why he traveled twenty-five hundred miles away from home to attend Colgate University in upstate

New York, in a culture very different from that of Seattle? Perhaps he needed to be in unfamiliar territory to really break away from his own home culture and become his own person.

As we got closer to Bubongo, my nervousness increased. Last time I visited the village, Hamisi was seventeen and living in a small homestead with his parents. Now he was a wise grandfather, and I couldn't quite picture what his home and life would be like.

As we began the gentle ascent toward Bubongo, more signs of civilization came into view, with small, thatched homes tucked into the foliage, surrounded by banana and cassava plants. Colorfully dressed women carrying water walked by; soon the red-brown mud-and-brick houses of the village and small pathways for foot traffic surrounded us.

The five-hour journey had not been easy in the heat, but I knew we would be stopping for refreshments. Abdul had planned a visit for us with Esilom, the field assistant who waited up for me when I was lost on the lake, on our way to Hamisi's. As we approached Esilom's homestead, my heart beat faster; then I caught a glimpse of him from afar. He had been a good friend as well as my field assistant on several chimp follows during my student days. Esilom's young face was imprinted in my mind, and I recalled his frequent shy smile.

"What do you think?" I asked Tommy.

"I can't imagine reuniting with someone I haven't seen for thirty-six years. You guys were single, carefree kids playing soccer on the beach the last time you were together."

Carefree—that was close to the truth, but only compared to my present life. Back then, life seemed loaded with self-concerns and worries—but yes, that had been compensated for by the spontaneity and fun in our lives. I wanted some of the carefree back in my life now.

We entered the small, earthen courtyard behind Esilom's house, where he and his extended family of ten welcomed us. Each person greeted us with, "*Karibu sana*," welcome to our home.

I was taken by the bright colors of the clothing the women were wearing and the gracious smiles on their faces, the squawking chickens running about, the clothes drying on the line, and the children at play. Tommy and I soon became the main attraction as other villagers I knew from the past stopped by to see us.

Esilom led us inside his home, which was constructed of mud and brick with a thatched roof and hard-packed dirt floors. The family had no running water or electricity, though there was piped-in fresh spring water available nearby. As we passed by the kitchen, I saw three women preparing a meal. Squatting on the floor, they heated food over a small fire and chopped vegetables just picked from the garden.

Esilom was quiet as he led us into the living room, where we sat on large comfortable chairs and sofas. He looked serious and was a bit slow in his movements. Of all the people I reunited with, he seemed to have changed the most. The playful, relaxed facial expressions I remembered had transformed into a wise but concerned look. I wondered if he had experienced difficult times.

Among his ten children, one was named Figan, after the chimp, and Figan was just Tommy's age. Figan told us, "When I was younger, I asked my father why he named me after a chimpanzee."

Figan said his father had replied, "I really like this chimp and have great respect for him, so I wanted to give one of my children his name."

A very articulate young man, Figan interpreted his dad's Swahili for us. It seemed strange calling him Figan, but I was impressed that Esilom had given one of his children the name of a Gombe chimpanzee. Like many of the field assistants, Esilom was proud to be part of the chimpanzee research and spoke of the chimps as though they were family.

After a small feast of very fresh chicken, garden vegetables, and rice, we gathered ourselves to walk a mile or so to Hamisi's house just outside the village. Along the way we met Hilali, Hamisi's older brother. Hilali had always carried himself proudly, and now he posed with family members with the same regal expression and posture as I snapped his picture. Disturbingly, I learned a few months later that Hilali had died rather suddenly. The cause was unclear but seemed to be some kind of infection.

Home Again with Hamisi

When we finally arrived at Hamisi Matama's homestead, he came right out to greet us, then gathered his two wives, many of his fourteen children, and his grandchildren, and welcomed us into his home.

As is customary in Muslim tradition in many places, the men sat in a circle on hand-woven carpets on the dining room floor for the main meal while the women prepared and served the meal and ate later. The children entertained themselves outside or in other areas of the house, but showed strong interest in us from afar. At times I would look up at the doorway or windows and see many heads peering in at us. They were quiet and seemed very curious but also content just to watch us.

Several large pots were passed around, and we served ourselves. One dish had piping-hot chicken and vegetables in a sweet, mild sauce. The cassava had been cooked over a fire and pounded until soft. It resembled mashed potato and took the place of rice. The other dishes included different types of vegetables in mild light sauces with tasty herb accents; only one had a spicy kick to it. Just as I was starting to wonder if the water would be safe to drink, Hamisi came back from the kitchen with a jug of water and announced to us that he had boiled the water. I was relieved and much impressed with his hospitality.

We remained seated on the mats to talk about our lives and hear updates on what other field assistants I had known were doing.

"Rugema and Yassini have passed," Hamisi told me in Swahili, soon after rejoining the circle with the drinking water. Everyone was looking down at their food but nodding their heads and making sounds of agreement as individuals contributed to the conversation.

I understood much of what he said to me, but it was helpful to have Abdul to translate whenever I turned to him with a confused look. Learning that two of the field assistants had died greatly saddened me. I saw their faces in my mind and couldn't comprehend their dying at forty or forty-five years of age. One of them, a good friend to me during my student days, had died of AIDS, and the other had apparently died from an infection.

My visceral reaction to these "early" deaths was influenced by my own culture in the States, where the current life expectancy for men is seventy-six and for women, eighty-one. In Tanzania, the life expectancy for men is fifty-one and for women, fifty-four. People can expect a quarter-century more life in my culture than in Hamisi's. Were I the average Tanzanian male, I would already have been dead for seven years. I was shocked and very disturbed when a patient in my practice died at forty-six from lymphoma. At the time he died, this was the average life expectancy for men in Tanzania, yet it seemed to me he was only halfway through his life.

The life expectancy for chimpanzees (once adulthood is reached) is between thirty-five and forty years in the wild, only slightly less than the forty-three year life expectancy for humans in Tanzania when Jane arrived in 1960. Recognizing this view of life from a cross-cultural and cross-species perspective helped me understand some of the terrible losses of people I respected and befriended at Gombe.

After our meal, we went outside, where Hamisi's family usually spent most of the day. One of his grandchildren, who looked to be about five years old, had physical disabilities and appeared unable to walk. As he sat on the ground in diapers, playing and smiling, Hamisi asked if I would take a picture of him. He was clearly very proud of the little boy, who seemed well integrated into his community. People helped him and played with him throughout our visit.

We spent the rest of the afternoon mingling with family members and villagers near Hamisi's house. I noticed Tommy talking and laughing with one of Hamisi's teen daughters, who had an unimaginably beautiful smile and quiet sophistication about her. As we posed for large-group pictures, I found myself placing my arms around grown women such as Hamisi's wife and sister and feeling a closeness in doing so. Kids were home from school by this time, and because the event had been planned ahead of time, family members were home and not working. Tommy said afterward that he loved seeing the women braiding each other's hair and the closeness of people sitting and talking in the courtyards. It seemed to him, and to me, that it would be hard for someone to feel lonely in a community that embraced its members as they did. (We did not delve deep enough to know if there were exclusions or prejudices in village life here.)

Groups of women engaged in conversation as they relaxed in the afternoon sun on mats outside. Infants strapped to their mothers' backs gazed at the activities around them, and older children laughed and played nearby. Their connection to each other and their connection to nature were strong. Their Muslim or Christian religion added another bond and sense of purpose among the villagers, and I heard of no conflicts with the two religions coexisting.

One-on-One with Hamisi

In late afternoon, I gave Abdul the time-to-go nod. I embraced Hamisi's family members and we set out so we could return to camp before dark. Hamisi volunteered to escort us to a shortcut, which took us an hour to reach. This gave us time for an intimate talk, which I'd hoped would happen during this visit.

I asked about his children, and Hamisi spoke at length about them. Then he suddenly stared at the ground and said in a hesitant voice, "The most unforgettable event in my life was the death of my son, who died in 1994, when he was eighteen years old. My life stopped for one year after he died."

It was almost expected in Tanzania that infants might die—the country's infant mortality rate was over 12 percent during my student days there. Out of eighteen births with two different wives, Hamisi had lost three infants, which was not rare in Tanzania at that time, but his eighteen-year-old son's death had impacted him severely. He was lost in grief. He could not understand why this death happened.

He went on: "He died of symptoms that were similar to cholera, but I don't believe it was cholera. I believe it was black magic, or voodoo, that was done against him. I loved my son so much because he dedicated his time to attending high school and also to farming. I believed he was the one who would have come to help me in the future, but it was not my luck. This is the only event that has happened that I will never forget for the rest of my life."

Trying hard to absorb the intensity of Hamisi's struggle with this loss, I thought about my childhood best friend's mother. Esther was an unfailingly devoted mother to her only son, Alex, who was just twenty-one when he died in an automobile accident. He was driving a tall, open-doored delivery truck that tipped over when a car ran into it, killing him instantly. He had planned to quit the job the week before the accident but had kindly stayed on the extra week because another employee was sick. Esther was so devastated that she could not adjust to life without him. When I appeared each year on her doorstep to visit her, she would burst into uncontrollable tears.

Then, twenty-some years later, she came to the door with a smile. She told me about a dream she'd had that allowed her to move on with her life. In the dream, God made it clear that he needed a drummer in heaven. Alex had been a talented drummer. The dream helped Esther reconcile the tragic loss with her religious beliefs. Her life was never the same, but she managed to feel better over time with a new image of Alex in heaven playing the drums for God. Her attendance at Catholic mass also helped her to maintain a spiritual bond with her son.

Hamisi had never found solace after losing his son over a decade prior. And I couldn't imagine how I would find peace with such a loss. It was a reminder to me of Flint, and of how loss can lacerate so deep that life itself loses meaning, or in Flint's case ends.

Tommy was walking right behind me, and I was overcome, realizing how lucky I was to have him and Patrick, my healthy sons, likely to live into their eighties or nineties. I couldn't imagine losing either of them. Amid my grief, I wondered if Hamisi's son might have had a different outcome with access to the diagnostic and treatment resources of a more developed country.

I wanted to communicate to Hamisi how deeply I cared about this loss in his life. "*Pole sana*," very sorry, was the best I could do at that time, but I also later asked Abdul to communicate my deepest sympathy to him. The message I sent read:

> I can't imagine how painful and confusing it must have been for you to have your son die in his prime at age eighteen. I hope you eventually find peace with this as the years go on. I knew from our time tracking chimps together in our early days and the guidance you provided me that you would make an excellent father, and I'm sure your son benefitted from that.

As we continued our walk, Hamisi said, "Now I want to hear about your life," looking more directly into my eyes than usual. I was quiet, finding it difficult to launch into the past thirty-six years, knowing he might not relate to any of it. I chose to describe my work as a family doctor and my kayaking in Puget Sound, and to tell him about the beautiful mountains and lakes in the Seattle area, which I thought might resonate with him.

I would have like to have talked with Hamisi a lot longer and heard more about his life, but the walk to the shortcut was coming to an end. Hamisi had hiked five hours to Gombe to greet Tommy and me on our arrival, and had planned extensively for our visit. My heart overflowed with gratitude for his generosity of spirit and for his friendship. I realized why we had become good friends at Gombe. I could still feel that same connection, and my response to his compassionate nature. I was filled with a sense of comfort as I recognized the completion of this part of our journey, a treasured reunion with a friend and mentor from a distant chapter in my life.

We parted with Hamisi on the path; our group headed back to camp and he to his village. I turned around and saw him, slightly lower on the slope, standing tall and serene on the dirt pathway with his bright blue-and-white dress shirt framed by the rich green forest plants as he watched us leave. My eyes were fixed on his. I didn't know when we would see each other again, if ever, and that was painful. But I felt that I had connected more deeply with him on this trip, and our bond was stronger than ever. As always, I saw that glimmer of a smile on his face. I remembered him exactly this way at age seventeen, waiting for me at the upper camp in the early-morning hours with our check-sheets in hand, before we went to tackle yet another adventurous day following chimps through the forest.

I had some photos and video of our time together, and many memories of our days following the chimps, but as I trekked back over the mountains, I realized that I still knew relatively few details about Hamisi's life, both past and present. It was too late to ask him any more questions, and I knew my Swahili wasn't advanced enough to understand some of the complicated emotions and events of his life over the previous decades. When we arrived back at our beach dwelling, I wrote out some questions and asked Abdul if he would interview Hamisi after Tommy and I returned to Seattle and then send me Hamisi's responses.

In my mind I carry an image of Hamisi, the elder wise man, being interviewed by Abdul, the learning young man, as Hamisi answers my questions about his views on life in the forest and in his village. I picture him thinking intently about how his life unfolded, and I'm grateful to have this information. Still, I wish I had been fluent enough in Swahili to ask him myself.

One question Hamisi answered involved his knowledge of both plants and animals, something I wasn't fully aware of during my student days. Hamisi wrote:

> My research work at Gombe helped me to learn a lot about the plants that are useful as local medicine to cure diseases like typhoid and malaria, serve as antivenom for snakebites, and treat ulcers and many more conditions. Also, I know different species of animals and their behaviors, what they eat, and how they communicate with one another, and can identify them by hearing their calls, especially chimpanzees.

I had been a beneficiary of Hamisi's many skills. I learned from the way he moved, analyzed his surroundings, and communicated his knowledge to others. Above all, I appreciated that Hamisi always seemed to be watching out for me without being overbearing.

"*Kuja karibu*," come near, he would say and quietly point to Fifi, who was selecting a perfect twig to use for termiting. Hamisi would then step back for me to continue observing the behavior that was crucial for me to witness. Back then I could picture guiding my future children this way, though I'm not sure I achieved the same level of confidence and competence with Tommy and Patrick as Hamisi did with me and, I presume, with his kids. It was sixteen years after leaving Gombe and after my first son was born that I found myself evoking this image of Hamisi gently allowing me to step out in the field while still supporting me from a distance. When my boys needed to step out into life on thier own, I tried to blend into the background as Hamisi did with me. This did not come naturally to me having grown up with a father who would usually take control of the situation at hand.

And always, in our adventures, Hamisi's mild manner and slow, methodical approach helped guide me through the forest. His sensitivity to others and alertness to their signals taught me the importance of not rushing. An instructive Swahili proverb is "*Haraka haraka haina baracka*," which means, "He who rushes has bad luck." Not rushing has been one of the hardest lessons for me to learn in life. I try, both as a doctor and a parent, to take the time to pay attention, observe, and think things through. I'm not always successful.

Another of Hamisi's responses during the interview highlighted our similarities and why we became close friends:

> Sometimes when I went to the forest with other observers I wished I could go back to the days in which it was just the two of us, John and me. My manners and personality, including patience, a strong work ethic, and trustworthiness, were similar to John's, and for this reason we grew to like each other and become friends.

I was humbled by the fact that Hamisi chose those particular characteristics to appreciate. During the interview, Hamisi told Abdul he was a bit nervous answering the questions, perhaps because he knew his responses might be included in a book, but Abdul and Hamisi still managed to produce five pages of insightful writing that I received in Seattle with great appreciation.

The path continues for Hamisi and me today as we live our very different yet interwoven lives far away from each other. Now we have met some of each other's children after having shared a unique adventure as young men starting out in life. If our strong bond and memories of each other could last thirty-six years, I'm certain they will keep us connected for the rest of our lives.

CHAPTER TWENTY

MORE FOREST REFLECTIONS: BROADENING MY PERSPECTIVE ON FAMILY MEDICINE

After passing over the mountain range for a second time that day, I stopped talking. I was too exhausted to both speak and carry on walking. My legs were growing weak and I felt short of breath as we began our descent. Eventually I noticed that my hiking companions had grown quiet too. The heat lifted the rich, earthy smells of dry grass and ancient soil. We could hear the late-afternoon baboon grunts, pant-hoots from chimps in the distance, and our own heavy breathing as we made our way toward camp.

I kept thinking about one of the older villagers who had approached me at Hamisi's house. He was a stranger to me, probably one of Hamisi's relatives, and he had slowly limped up to me. He had a weathered face and a wiry build, and wore a thin blanket wrapped around him. He looked into my eyes, and with a kind and confident voice, said in Swahili, "Can you get me some pain pills for my arthritis?" He was referring to his hip. Apparently the local liniments he used were not effective for this degree of pain, due to what I assumed was a worn-out hip joint.

Walking along and looking at the peaceful, golden countryside, I couldn't help but contrast Hamisi's life as a field assistant and medicinal plant specialist with mine as a doctor in a bustling American metropolis, as well as compare the kinds of medical care available in each of our communities. There's a huge gap between a Tanzanian man asking for pain pills and the world of Western medicine, where he would have simply been offered a hip replacement for his advanced arthritic condition. The man's life might have been quite different if he had had a new hip—and my life might have been different had I chosen to provide health care to an African village. There are so many ways to improve the quality of human life—from the most cutting edge of life-saving medical interventions to the simplicity of relieving pain. In a perfect world, a health care provider could improve life wherever he or she encounters it.

I suddenly thought I might have been more satisfied as a Tanzanian medicine man than as a Western doctor. After hearing about Hamisi's personal losses as well as the many other health issues in Africa, I found myself wondering what would have happened if I'd returned to Tanzania to work after completing my medical training instead of staying in the United States. Would I have been more fulfilled working in Africa? Would I have done more good there than in a country with many doctors? These thoughts preoccupied me as Tommy and I hiked with Abdul and Rudo back to Gombe from Bubongo Village.

For as long as I can remember, I've been attracted to cultures different than my own. Perhaps I was searching for more inclusive and accepting social environments that would accommodate my shy and unaggressive nature. Whatever the reason, I continue to enjoy learning about other cultures. In particular, I'm drawn to learning about cross-cultural health beliefs and spiritual practices that influence people's lives across the globe. It's been especially enjoyable to see how my boys adapted in diverse settings as Wendy and I made choices that introduced them to different cultures. This included our trip to Gombe as well as a two-month stint the family spent in Barbuda, near Antigua, where I practiced medicine in a setting that reminded me of some of the towns in Tanzania. I also brought my family along when I signed up to serve as the ship physician for Semester at Sea, a seagoing university for 750 college students, which exposed our boys to Indian orphanages and Cuban medical centers, among other interesting sights in the four-month voyage around the world. These trips broadened my children's perspectives, but also helped me expand my view of what it means to practice medicine in diverse communities.

The Highs and Lows of Modern-Day Family Medicine

Throughout my long career in family medicine I've had definite highs and equally definite lows. Mid-career, my health care organization implemented electronic medical records and also increased the number of patients in our practices considerably. So began the dark years. It didn't help that I was experiencing sciatic nerve pain in my leg from repetitive trauma to my back from playing league soccer in my midforties. Sitting was difficult. Though this should have given me more empathy with my patients' aches and pains, at times I felt like saying, "You think *you* have pain!"

A period of five frustrating years resulted for most of us in family practice including primary care physicians, physician assistants, and nurse practitioners across the country as demands rose and resources didn't. Twelve hour days of relentless concentration on patient care took over our lives; I had to learn the new computer system, email patients, and perform other virtual visits such as phone appointments, in addition to seeing twenty-two patients a day. Applications for residency positions in family medicine plummeted because of the unsustainable workload throughout the field of primary care.

I remember sitting at my computer one evening at eleven P.M., reading an email from a patient. This was after a busy day seeing patients from 8:00 A.M. to 6:00 P.M. I had just completed looking at twenty lab results and responding to six nursing questions about advice for patients.

My brain was clogged with too much information as I read my patient's message about headaches that started near her left ear, then radiated to the top of her head, followed by a buzzing sound and possibly decreased hearing on the left. She mentioned that her jaw was also bothering her. The headaches had been going on for a week, and she wanted to know what I suggested. Headaches such as she described can be due to stress or tension, but they can also be due to a brain tumor called an acoustic neuroma, or perhaps temporal arteritis, which if not promptly treated causes blindness.

I stared blankly at the email for five minutes. We call this "brain freeze." I was too tired to think or get angry with my computer, my superiors, or the world, so I turned and gazed at the photo of Bubongo Village on the wall and the picture of several chimps feeding on milk apples as I tried to escape the reality on my computer screen. As I looked at the images, I felt a familiar sense of grounding and calm flow down my neck, which helped reenergize me.

When I did come back to the task at hand, I had enough focus to make a clinical decision and type it as an email to my patient. The next day I was relieved to discover that the woman had neither of the serious disorders.

I began to toy with the idea of leaving the practice because of the stress it was placing on my family and me. My anger with the untenable workload was building. A big frustration even today is how my colleagues and I care about the people we treat and are willing to go the extra mile for them, but at the end of the day, many of us are burned out. We've lost some of the joy and energy necessary to feel good about our work and about ourselves. We trained to be healers as well as clinicians, and modern medical care doesn't always value this distinction. There's little breathing room or processing time during a typical day unless you give up the time it takes to really connect with patients and simply respond, like a computer, with mechanical input and output—and that is not doctoring in a human or healing way.

Then came the "medical home," which allows more time with patients for each visit and time for emailing and phoning them about their health problems. A collaborative approach to diabetes, heart disease, and other chronic conditions by nurses, pharmacists, medical assistants, physician assistants, nurse practitioners, and family physicians with individualized care plans resulted in better outcomes for patients. We hired more physicians to allow for manageable practice sizes. The days are still long, and most of us email our patients from home, but the hectic pace has been reduced, as a portion of our computer work has also been delegated more appropriately to nursing staff.

Along with my complaints, I should acknowledge the bigger picture regarding medical care worldwide. The one doctor to twenty-six hundred patients in my practice during the dark years might have seemed overwhelming, but not compared to the one doctor for fifteen thousand patients in areas of northeast Africa and many other parts of the world. A family doctor I know from college days recently described his experience in Dar es Salaam; he would see up to two hundred patients a day in a clinic there, working with six medical assistants and two nurses as support staff. He had time only to hear patients' brief history and then to treat their infections or injuries with speedy efficiency.

Compensating for some of the frustrations and long hours in family medicine is the joy in delivering babies. This process is rewarding and usually exciting, though we always hope not too exciting. I chose not to continue the obstetrical part of my practice eight years ago, but I felt great satisfaction delivering babies and growing closer to the families as we spent hours together during the labor and delivery.

I'll always remember the delivery of a newborn whose two siblings watched the birth. In my practice, I tried to accommodate special requests of parents for the births of their children and approved the presence of this couple's three-year-old son and five-year-old daughter in the birthing room.

Though the children were very young, the parents were prepared to explain things as they went along. The birthing room was large and comfortable—though being in labor is never comfortable—with a rocking chair and Jacuzzi. This mother was a pro at this point, having gone through labor twice before. Her labor went fast, and as the baby's head began to emerge, everyone gathered close to see the new arrival. After the head eased out, there seemed to be a halt; the body didn't follow as expected. One of the most stressful emergencies in childbirth is a shoulder dystocia, in which the shoulders hold up the delivery. If the infant is not delivered within a few minutes, the brain can be deprived of oxygen, leading to brain damage or even death.

As I applied traction to the head, the nurse applied firm pressure on the mother's lower abdomen above the pubic bone to push the infant's shoulder down into the birth canal. I had no idea what the siblings were thinking, but when I glanced up at them, they both looked bored. Luckily, they had no clue as to the seriousness of the situation. I calmly but determinedly asked an attendant to page the obstetrician on call STAT. I continued to use a calm voice, though my face was flushed and my heart racing as I reached in to adjust the position of the baby. This finally allowed the big guy to get unstuck and emerge.

After a brief pause, the infant cried; then the mom started crying, and then the dad. The couple embraced as the baby, looking strong and healthy, nestled into his mom's breast.

The oldest child said, "That was cool." This was her only comment after the dramatic, nearly heart-stopping delivery.

Both children came over to look at their new brother as I attended to the placenta and my tension eased. I think the kids held up much better than the adults in the delivery room had and likely benefited from seeing the birth, which had been demystified in a way by their inclusion in the process.

As my time with the chimps connected me closely with the family life of the mother-infant pairs I studied, delivering babies in my practice allowed me to feel a similar closeness to the family dynamics of my obstetrical patients and their families while they toiled through labor and happily welcomed the new member into their family. For me, there was down time—sometimes *lots*—to ponder life and take in all the emotions that were flying around the delivery setting. And there was no set schedule to follow, just Mother Nature. Thank goodness!

The Turbulent Teens

It was a bad joke: "Looks like I'll have time to get caught up with paperwork this afternoon," my colleague blurted out to me early in my career. "I have two teen exams in a row." The somewhat humorous but very poignant comment referred to the fact that teenagers in general don't reveal much or talk much during the exam. The aloof gaze, the dangling hair over the eyes sometimes hiding acne, and the brief answers—all common features of the teenage exam. Parents were of course politely asked to leave the room for a short time so that these developing adults could unleash their opinions about their parents and reveal other secrets in a confidential setting.

Despite the almost comical difficulty involved in getting a good history from teens, I eventually enjoyed learning about and even delving into this scary area of development. After all, I had seen my boys traverse this phase of life, walked through it with my earlier patients, and witnessed the tumultuous time in adolescent chimps when I was a student at Gombe.

I became curious about the purpose of this volatile period of life in terms of human and nonhuman-primate survival. Accompanied by the impulsive and sometimes high-risk behavior of teens, there is an important "plasticity" of the brain from puberty to the end of adolescence; plasticity refers to the brain's ability to change and adapt to the specific environment in which we are raised. This ability of the brain to adapt, learn, and even physically change during adolescence was likely crucial to humans' ability to adapt to highly variable environments throughout our history. Because of a less sophisticated cerebral cortex in chimps, their plasticity is not as strong as ours during the teen years, and yet it's still important. Both species are programmed to leave or detach from their caretakers and are attracted to novel and exciting situations.

As I attempted to reach the hearts of my teen patients and my own boys during this stage of development, I really had to summon my mantra, *Remember Fifi,* a lot of the time. Her never-ending patience with energetic and sometimes wild Freud stayed glued in my mind. In the end, Freud turned out fine and continued to have a close bond with his mother. And it really does take a village in some teens to manage this stage in our current electronic world of cell phones and computers. From DBT (dialectic behavioral therapy) groups for kids in need of peer support to total hands-off treatment for the high achievers, there is always hope that by age twenty-five, when the human brain is fully mature (twenty-two in females), there will be a return to a normal and more peaceful existence. I am hopeful that more emphasis on adolescent health will result in better outcomes for teens in our society and throughout the world.

Lessons from Traditional Healing

Just as in Western countries, medicine in Tanzania has evolved with changing demands, but many people still view the traditional village medicine man as an important health care provider. When I was a student at Gombe, the local medicine included strong elements of holistic care. People living in rural villages were most comfortable with very personal, individualized care from a medicine man, and were willing to pay a fee for this service even though there were free government clinics in most nearby cities. This tradition continues today, though not to the same degree.

Recently, I came across an article I had saved from the April 1974 issue of *Hospital Practice* that summarized health care in Tanzania at that time. A quote describing a typical patient caught my eye: "He is likely to object to a crisp treatment of his symptoms, when he has been accustomed to the holistic approach of the medicine man for whom the patient exists as an entity with no clear demarcation between psyche and soma [mind and body]."

The patient's social surroundings and even the spirits of those in the community who have died are considered. Treatment often includes herbs and other plants from the environment. As the healer assesses the patient's problem, he continually looks for feedback from the patient as he takes a history and begins to give advice for the problem during this initial assessment. If the patient feels that the healer is off track, the healer graciously refers him or her to another healer.

Early in my practice, just after completing residency, I offered counseling sessions to patients enduring difficult times. A female therapist and I met with couples having relationship problems to give them support.

I enjoyed having thirty-minute appointments to delve into their marital lives as we learned about their specific issues. As we offered advice, we also tried to model good communication between the two of us performing the counseling. This opportunity soon went by the wayside as demands for more regular fifteen-minute office visits took priority. I missed this deeper connection with couples, a holistic approach that recalled the kind of traditional healing I had seen in Tanzania, where all aspects of people's lives are considered in their treatment.

Over the years, I've also found it fascinating and instructive to tease out a patient's own view of his or her health challenge. One of my senior instructors in medicine taught me this early on. During one rushed clinic day as a resident, I hurried to this preceptor's office and presented my required history and assessment of a young man with a sore throat whom I had just examined.

The preceptor, as usual, said, "What does the patient think is going on?"

I refrained from saying, "It's a simple sore throat! Why do we need to know what the patient thinks is going on?"

Instead I went back to the exam room and asked the patient what he thought was causing his sore throat. He immediately opened up about his fears after having engaged in oral sex for the first time and voiced that he wondered if the throat pain was related to this. He was clearly anxious in general about the encounter and thought he might have contracted a serious disease, but he wasn't able to bring up his concerns without my directed questions. I learned a lot from this and—I hope—helped the patient, who would have gone home still worried, his concerns not aired, if I had not invited him to share them. I could hear his sigh of relief as I explained the low likelihood of any serious health consequence in his situation, and I did a throat culture for further reassurance.

Once again, through creating a safe space for a patient to talk and through listening and asking questions, I was able to be both a doctor and a healer. As my medical training continued, I found that it was always important to try to be both. Doctoring may be something one does in the moment of diagnosis and prescription with a patient, but healing is something that needs to be addressed by taking a broader view of the patient's well-being over the long term. Whether it is a broken ankle or a breakup with a close partner, the healing can take months or years, and I found that gentle reassurance along the way could be very helpful. The same gentle reassurance that was the hallmark of Fifi's mothering—not overdone, but steady and available.

Walking back to Gombe with Tommy, Abdul, and Rudo, I considered the possibility of spending time in Bubongo sometime in the future to learn more about the villagers' health beliefs and their ways of dealing with illness. I was still curious about how healers in diverse cultures assessed the physical, psychological, and spiritual needs of their patients. It must have been frustrating to work as a doctor in Bubongo, as in other parts of the world, having few resources for preventing infections such as malaria and dysentery, especially in children, yet I was sure it was tremendously satisfying to help people in such dire need of care.

I was also sure that there was a mutual benefit in sharing the lessons of Western medicine and those of traditional indigenous healers of East Africa and the rest of the world. The medical home model being developed in my Seattle practice includes psychosocial elements similar to those used by traditional Tanzanian healers, looking at illness in relation to a patient's social surroundings and family. But it also utilizes a state-of-the-art electronic medical record system to track information and allow patients to communicate online with their providers. I envisioned a future for family medicine in which the best of both medical worlds—traditional indigenous and modern Western—would be merged. Our learning goes on.

CHAPTER TWENTY-ONE

GOOD-BYE AGAIN

On our last day at Gombe, Tommy and I awoke later than usual, tired after our trek to Bubongo. By late afternoon we would be in Kigoma, and then on to Zanzibar before returning to Seattle. "What do you want to do on our last morning here?" I asked Tommy.

"Not much," he answered, almost falling back asleep.

We spent the morning of our departure down by the lake, saying good-bye to people and taking our last swim. Small boats passed by, women washed clothes in the lake, and people strolled along the beach. The lake had been crucial to my tolerating the hot days of the dry season when I was a student and also provided me with memories of brilliant light spectacles in the sky as the sun disappeared beyond the horizon. Somehow the perilous side of the lake did not remain strong in my vision of the large body of water. Besides the occasional water cobra that would swim close to me at times, I also encountered a larger creature during one of my early evening swims. Having completed my tracking of Melissa and Gremlin, I swam out from shore farther than usual. Looking back at the shore I saw a group of Tanzanian men and women waving to me, and I waved back. As I noticed their waving becoming frantic, I looked away from the beach and toward the west to see a hippo staring at me from about fifty feet away. Hippos are known for being highly aggressive and are ranked among the most dangerous animals in the world. In the most adrenaline-fueled swim of my life, I pulled hard at the water without pausing to look back. I reached the shore in time for the larger group who had gathered to break into applause as they waded out and hauled me safely to the beach. This memory arose now but without the terror; I mostly remembered the support I received from the villagers at the camp.

I sat down in the sand by myself and leaned up against some rocks to view the beach activity and expansive lake. The chimps never enter this territory, so with my back to the forest I felt that I had already departed from their world. But what I really wanted was to have a few more hours in the forest with Freud and Frodo and to find Gremlin, who was now the mother of twins. I wanted to watch them all one last time. My reconnection with the chimps had been fulfilling, but I wished I'd had more time with them.

Looking down the beach, I pictured Hamisi walking toward me as he did when I stepped out of the boat, having not seen him for thirty-six years. Our connection seemed to transcend race, geography, and wealth. We each felt we shared similar personality traits, responses to our environment, and views of the world. I wondered what it would be like for Hamisi to visit my home in Seattle. From his cautious reaction to our past trip to Dar es Salaam and Kilimanjaro, I'm not sure how he would respond to the tall skyscrapers and intense traffic of our bustling city. Perhaps a kayaking trip in the San Juan Islands would be more appreciated.

A small wooden boat with a ragged cloth sail breezed by, holding a family bringing vegetables back from the market to a village north of Gombe. When they caught sight of me they flashed big smiles, and I called out, "*Hujambo*," hello. Just as I had pictured spending the night in Bubongo Village, I imagined spending the day with this family as they sailed to Kigoma and back with their bountiful produce packed in their small boat. I wondered what brought them joy and what stimulated them to carry out their lives the way they did.

I thought about Tony, who had made Gombe his permanent workplace. I walked by his house on the beach, made of aluminum siding and thatched roofing with large screened windows open to the lake breeze, and realized that, although I didn't choose that path, it would have been an amazing life to be so closely connected to nature. I acknowledged to myself the admiration I felt for Tony and how happy I was that he had assumed a major role in managing the work at Gombe.

I walked by Jane's simple house too. Stopping, I imagined her running out of the house, leaping into the lake, and swimming out to get the wrapped rose dropped from the sky by Derek, her husband-to-be. I pictured Grub at age seven, playing on the beach. Staring at her former dwelling, which now serves as a guesthouse for visitors, I thought, *So much has changed since 1973.* The people and chimps had grown much older, of course, living out their lifetimes in some cases, and Jane no longer lived here year-round. It was reassuring to see that the chimps were thriving and that the research was continuing, but I felt like an outsider now and longed to be a part of it again.

To get a break from the bright sun, I wandered fifty feet into the forest and found a small knoll where I could pause and feel more connected to the chimps before leaving. Tommy remained near the beach visiting with the staff. There was a perfect, natural seat scooped into the ground, surrounded by amber-colored grasses shooting up from the red-brown earth among twelve-foot leafy trees with bright green leaves and dangling vines. Although far from Jane's Peak, I thought I would privately name this John's Peak. I concluded in my spiritual imaginings that this would make it more likely that I would return to Gombe someday, perhaps with Wendy or Patrick.

As usual, the setting allowed me to focus on natural elements around me and not think so much about my worries and regrets in life. Sure I wished I had learned Spanish, spent more time with my kids just having fun, worked as a doctor in Tanzania, and developed musical talents. But at this moment of reflection, alone on the small bluff in equatorial Africa, I let go of these regrets. There was a bigger picture surrounding me physically and emotionally. As in my student days, I embraced a different perspective. I had fewer expectations of myself in this big, natural world around me. Although I appreciated Jane's message, that each person can make a difference in the world, I felt less pressure to be or act a certain way. My feeling of purpose and motivation emerged from a stronger sense of myself and what I could contribute to society. Nature had a way of clearing the clutter for me and adding clarity, like the visual separation of pure white mountain snow against the deep blue sky on a clear day. I was sure Jane felt similar peace and connection in Gombe and was extremely torn when she made the decision to leave her forest home in the late eighties to begin her indomitable crusade across the globe to save species and land, including chimps in the wild. Jane still visits Gombe a few weeks each year, and several of the field assistants told me quite confidently that the chimps know when she is around and show up in or close to camp soon after she arrives.

After an hour of relaxing and thinking about my past days as a student here, I rose up, dusted off the dirt, and headed down to get Tommy for our departure. At two in the afternoon, when the boat was ready to leave, whitecaps spread over the lake. Tommy and I had packed our clothes, eaten a leisurely lunch, and unwisely chosen the afternoon to go back to Kigoma. I'd forgotten that the wind always picked up in the afternoon. I recalled a similar scene from my student days here, when the boat carrying prominent English judge Lord Denning and his wife Lady Denning, in their seventies, had to be lifted by sturdy Tanzanian men over breaking waves in strong winds to begin their return trip home to England. I had hoped our departure would be less dramatic.

Abdul arrived to see us off as our boat driver readied the vessel. We smiled and shook hands, but our good-byes felt a bit rushed with the strengthening wind and the boat rocking in the waves. "You must come back here again soon, Bwana John," Abdul said, as we instinctively hugged each other. Tommy smiled and made sure we had all our belongings. "*Kwaheri*," good-bye, he said as he jumped in the rocking boat. Twenty feet from us on the shore a park warden held a rifle in his arms, a shocking sight to me on this return visit. That sight would have been unheard of thirty-six years earlier when I was there. With the 1974 kidnappings, increased poaching in other areas of Tanzania, and terrorism on the continent, some of the park wardens were now armed. I guessed if it meant protecting the chimps and researchers, it was needed. My eyes quickly returned to Abdul as he stood at the dock and watched the boat pull away. He kept watching us until we were out of sight.

I looked back at the lush valleys and undisturbed beaches and wondered how they'd look in ten or twenty years. Would Gombe survive all the changes to come? Would Jane's work and others' ensure that the chimpanzees would thrive in this ancient forest for another century?

When I could no longer see these rugged valleys and the beaches I knew so well, my thoughts turned to the logistics of the journey home. Like a switch, my connection with the chimps was turned off. I relished my new connection with Tommy and how our common adventure to chimp-land might bond us further as we share the meaning of that trip into the future. But for now, the chimps seemed like they were on a different planet. I could see some of them online over the coming years but would not feel the close connection I had the privilege of experiencing on two wonderful trips to the Gombe forest. In two days I would be back in the office, seeing my patients and hearing their stories of their lives and concerns, very different from those of the Bubongo villagers. I would keep the vision of my time sitting on the knoll and also in the warm sand on the beach close to me, especially during stressful times and long rainy winters. I would picture Freud and Frodo relaxing on the hillside as the sun approached the horizon, munching on milk apples before climbing up a leafy tree to build their nightly nests.

CHAPTER TWENTY-TWO

KEEPING UP WITH JANE

It was hard not to think of Jane while at Gombe since she had been such a big part of that experience for me before and during my time there. And as I mentioned, we had kept in touch through letters, get-togethers, and memorable reunions—large and small—over the years.

The first reunion took place within a year of my leaving Gombe. By coincidence, another student who had worked with Jane, Chuck de Sieyes, was also a student at Case Western Medical School. The two of us enticed Jane to meet us at the Pink Pig, a rustic retreat center on beautiful farmland in Ohio as a break from her tight speaking schedule in the States.

It was 1975, and the ordeal of the internationally reported kidnapping of three Stanford students and Emilie Riss (whom I had worked with in the Gombe medical clinic) had just ended. The students had been taken by boat from Gombe and held hostage for two months by Congolese rebels. Very tense negotiations with the rebels had transpired in order to finally secure their release.

Jane had been stressed and drained. I could see that the threats of closing down Gombe, ending her ties with Stanford, and worries about the students had taken a toll on her. But she still cheerfully said, "What an unusual and charming farm. I love seeing pink pig pictures on the walls throughout the center." It was so heartening to see her laugh and have fun with us in this quirky setting where everything, including the walls, tables, floors, and barn, had a pink pig motif.

"Jane, how are you holding up these days?" I asked.

"Right at this moment I feel reassured having you and Chuck at my side, thinking of the good old days at Gombe together, but the kidnapping has taken my energy away. I appreciate having you two spend this time with me."

We reminisced about Gombe, walked the farmland as though it were the African forest, and enjoyed our time together immensely. Even years later, Jane shared how important it had been for her to reconnect with her students and feel our love and support.

A few years later, Jane was speaking in New York City at the Museum of Natural History soon after another tumultuous time in her life. She had no idea I was passing through New York, but I bought tickets to her talk. Despite my reserved nature, when I arrived ten minutes before she was to be introduced, I asked an usher to please take me backstage to greet her and let her know I was there. When I saw her seated behind a massive curtain, gathering her thoughts for the talk, she looked distant and very alone. Derek, her second husband, had died of colon cancer six months earlier. Jane wore a beautiful formfitting dress subtly patterned with flowers that made her look serene and angelic, despite her sorrow.

A smile crossed her face when she saw me, though her eyes remained sad and tired. She gave me a huge hug, and when I complimented her on the dress, she said, "Derek wanted me to wear this."

Some years later, the thirty-year Gombe reunion at the University of Minnesota, where Jane's archives were kept, was an especially joyous occasion. Jane was full of energy, humor, and appreciation. As Wendy and I walked into a large reception area with her, she let out a perfect, very loud pant-hoot that echoed throughout the hall. Of course a resounding response of pant-hoots by most all of the people attending shook the walls. This set the stage for an evening of sharing memories of our Gombe days and getting caught up on one another's lives.

After dinner and Jane's warm and sentimental welcome, Emilie Riss took the stage and talked about life after her kidnapping. "As most of you know, my husband, David, proposed to me immediately after I was released. After barely surviving a car crash while driving in Tanzania soon after this, we married and have raised three beautiful children on our farm in New Hampshire. David continues to work as a family doctor, and with my background as a veterinary technician, I devote my time to the animals and of course the kids."

Tony, the baboon researcher, described his long commute from London, where his Tanzanian wife and son lived, to Gombe, some five thousand miles away, where he worked with the baboons. "This separation isn't easy, but I can't give up either of these aspects of my life," he acknowledged.

When Jane began her talk, I could tell by her nonstop smiling and her laughter that it was an emotional evening for her, to have not only her close friends there but also the team of researchers and former students who were dedicated to the Gombe chimps and to environmental preservation. For me, this great escape from my time-consuming medical practice spent reconnecting with many of the forty fellow researchers and administrators was very stimulating and fed deeply into the highly sentimental part of my soul.

"We are one big family," she said in a soft voice, "and we must not let another thirty years go by before our next reunion."

Nearly one-third of the nineteen Stanford students who were part of the Gombe "family" over the five years of the university's involvement became primary care physicians. We had all chosen human biology as our major, which included a mix of anthropology, sociology, and psychology instead of traditional biology or chemistry, and perhaps that selected for more generalists (none of us became neurosurgeons). We all were fortunate to have had a break from classroom learning to witness how our closest living relatives survive in the wild. We ventured into an African forest of discovery and learned from "jungle professors"—the chimps and Jane—about primates' basic needs. Both Jane and Fifi were superb instructors, and we all rated them quite highly.

Words from Other Former Students

Recently, I asked other physicians who had been student researchers at Gombe how the experience affected them in their careers as doctors, and three sent me their impressions. Because Jane, in her generous way, had included undergraduate students in her renowned research, each of these doctors had been molded by this extraordinary experience. Jane's impact on their work as physicians was significant.

For instance, Nancy Merrick, a practicing internist in Los Angeles, commented on how she came to better understand the genetic component of human behaviors:

> The Gombe chimpanzees taught me that much of who we
> are is to be found deep within our genes. Not all, mind you.
> Family and life experience also have played an important part
> in ensuring our well-being and values. But a vast proportion
> of the people we grow to be, I believe, was determined before
> our birth, encoded within our DNA.

Years before genetic testing was able to determine the fathers of the Gombe chimps, we often guessed as to the various sires. It seemed, for example, that the fairly even-tempered Evered surely must have sired Melissa's affable daughter Gremlin. Generations later, when Titan began to erupt into aggressive behavior, most everyone ventured that his father must surely be the belligerent Frodo. Each of these youngsters was reared by its mother, without the influence of the father, leading us to believe their likeness was inborn. In fact, genetic testing ultimately confirmed both.

Every day at the clinic, I see parts of my patients' natures that seem instinctive or predetermined, particularly the predisposition that some have to addiction or obsession that appears once amplified by life's circumstances. No amount of my counseling is apt to solve some of these medical issues. They run far too deep for my machinations at the surface to reach.

The chimps taught me, too, much of the biological basis of human behavior—our competitiveness, social hierarchies, compassion, need for reassurance and the touch of others, our love of laughter and playfulness. We share it all with chimpanzees and other intelligent animals. I hope I am somehow a more forgiving physician, knowing it is oh so difficult to overcome our most deep-seated selves despite our big human brains.

Chuck de Sieyes, my medical school friend and now a family physician practicing in Maine, shared his appreciation for a mentor and some other notable moments from his Gombe experience:

When I arrived in Gombe, I already thought of myself as something of a junior doctor. (Talk about being cocky!)

While in Nairobi waiting to fly into Kigoma with Jane's ex, Hugo, I ran into Dave Furnas, chief of plastic surgery from UC Irvine, who was doing a "Flying Doctor" stint. He was wonderful enough to take a real interest in me, so he invited me to come out into the bush to watch him operate. Since I was not yet a "real" MD, I couldn't fly in the famous Flying Doctor airplane (although I have a favorite photo of skinny me standing proudly next to the plane), so I would take buses at 4:00 A.M. with all the locals and their livestock in order to don rubber garden boots and scrub in at tent hospitals out in the country-side. He performed miracles with those skilled hands of his: rebuilding a child's face that had been mauled by a hyena; constructing eyelids for a baby born with no eyes and closed sockets, repairing cleft lips and palates. Talk about firing up my enthusiasm to get going with my *own* medical studies! I loved every minute of it!

So, with my ego flying high, I arrived at Gombe laden with a bag full of antimalarial drugs in hopes of crusading preventive health among the trackers and their families. Every week I would knock on the door of each and every one of the guides' quarters and hand out tablets to the entire household. Not only will I never know whether any of them actually took the pills (and if they did, it was more than likely to please this altruistic blond *Wazungo* [white person]), but I had a rude awakening about every six weeks when at certain doorsteps I would be met by a *new* wife—as is tribal/religious tradition for many of the trackers' families! Needless to say, I learned a thing or two about superimposing my Yankee value system and good intentions upon a societal structure I really knew nothing about.

Kathryn Morris, a family physician specializing in women's health care in Santa Cruz, California, shared a poignant lesson learned from a baboon mother and how it influenced her practice and her personal life:

> I have always been fascinated by evolution and how similar we are to other animals; my experience at Gombe really brought this home to me. As a family physician, I chose to focus in the areas where my interest in and understanding of evolution and animal behavior served me well. Delivering babies was a good example of this as it is so clearly a time when humans do best when in a nonhuman-animal mode. My interest in obstetrics was perhaps also augmented by my own personal fears about giving birth.
>
> I used to tell pregnant patients a story about a baboon that went into labor while I was at Gombe. All was going well until a leopard passed by, whereupon her contractions appeared to completely stop for a few hours. The baboon troop moved to a new location up in the trees, and evidently the contractions resumed. All went well as she had a healthy newborn clinging to her the next day.
>
> I told my patients this story as a reminder that our bodies know not to give birth in dangerous situations. It was fear that stopped labor for the baboon, and fear does the same for us. "Failure to progress" is the medical term used for labor that is not making good headway, and it is the most common reason for doing cesarean sections. It made sense to me that reducing a woman's fears could be key to her having a successful vaginal birth; clearly I was not the first to think of this, since midwives wrote whole books about it. Nonetheless, I did what I thought would build a woman's confidence in her body and help her feel safe and relaxed about the upcoming labor

and delivery. This usually included a good diet and lifestyle, birthing classes, and counseling if there were family issues. I often had women write a vision of how their labor and birth would go. During labor I felt it was important that all support people, particularly the medical staff and myself, remained calm and encouraging. I found that blood pressure checks and pelvic exams could be timed better so as not to be as disruptive, and monitoring equipment could safely be minimized. Allowing a woman more physical freedom, encouraging her to move around, go for walks, and take a shower all seemed to help labor progress.

It appeared to me that the more relaxed the woman was, the more likely her labor would proceed normally. Whether women with more normal labor felt more relaxed or more relaxed women tended to experience less problematic labor is hard to say for certain. However, my patients experienced a 6 percent cesarean-section rate, which was much lower than that of any of the other twenty-five doctors in the hospital where I did deliveries. All of these babies were healthy, and I count my blessings for this, since bad outcomes can happen even with the best of doctors.

When at the age of forty I was in labor at home with my one and only child, my moans and groans didn't scare me, as they often do laboring women and their inexperienced attendants. Instead, they were reassuring because they sounded like a healthy labor should. I was fortunate to have slipped my husband into several deliveries as the photographer, so he too was familiar and comfortable with the sounds and the ambiance of a woman in labor. All went well for us, and I gave birth to a perfect seven-pound twelve-ounce girl. My midwives said that it was an unusual birth for them. At first they weren't sure why, and pondered if it was because I was a doctor, but then they realized that it was because of my lack of fear. I felt much gratitude for the 650 women who wanted my assistance in the delivery of their babies, as it was through those experiences that I learned to get past my own fear of giving birth. And I thank the laboring baboon in Gombe who taught me that fear itself is the biggest hindrance.

Reading these responses, I was struck by how each of us was affected in such different ways by our Gombe experiences. A common thread was our interest in analyzing why we humans do the things we do and how our behaviors relate to those of other primates. As our mentor in the field, Jane encouraged all of us to keep thinking about the Gombe chimps and baboons long after we left. We were all motivated to find purpose and passion in whatever we did—Jane's way.

Beyond Jane's extraordinary talent as a mentor to scores of students, researchers, and others, her ability to keep up a marathon travel pace was another of her traits that astonished me. I know she became exhausted at times, traveling more than three hundred days a year. She spoke at schools and churches and for environmental groups, businesses, and dignitaries around the world. She used her time wisely and, at age eighty, had a memory and intellect that made for exceptional talks and discussions about endangered species, our planet Earth, and—always—the Gombe chimps.

In 2000, United Nations president Kofi Annan appointed Jane as ambassador for peace, and she committed to the role with her renowned energy and enthusiasm. She devoted herself to organizing events throughout the world to promote peace. Despite my disdain for assembling anything with printed instructions, I went to her website with then-five-year-old Patrick, and together we constructed the featured twenty-foot white dove out of old sheets and recycled chicken wire. On a beautiful September day, which had been declared World Peace Day, Patrick and I and a small group paraded the bird high in the air around Green Lake Park in Seattle, singing peace songs from the sixties as people stared and smiled at us.

Jane's gift for engaging graciously with others always extended beyond her public life. I appreciated Jane's interest in and interactions with our boys, so in keeping with her nurturing and caring nature. Watching her instigate a vigorous pillow fight and tickling game with Patrick when he was a toddler and talking interestedly to Tommy at different stages of his life meant the world to Wendy and me. Every time my family and I attend one of Jane's talks in Seattle, she always introduced us to the audience—as she did her other friends in attendance—a way of acknowledging us all.

It was never surprising to see prolonged standing ovations both before and after Jane spoke from the packed crowds who came to hear her presentations. Her impact is inspirational. The biggest hallmark of her style in this setting is her ability to draw on memorable examples to illustrate her points. During one appearance, she raised a clear glass of water to illustrate how fortunate we were to be among the one-third of the world's population to have access to clean drinking water, something we took for granted. I still think of her holding that glass when I drink Seattle water from the tap. Whether asking a young child from a Roots & Shoots club to come up onstage, or holding up a huge condor feather given to her by a group who had succeeded in keeping the species from going extinct, Jane motivated her audience to go out and do something more to preserve wild species and the environment.

Jane has a gift for empowering people around her. At one Seattle get-together on a beautiful spring evening a few years ago, thirty people gathered with her in a host's home near the airport for casual conversation and a vegetarian dinner. After our meal, Jane quietly organized us into a large circle to begin a talk. Our group of naturalists, journalists, friends, and educators suddenly had a purpose beyond dinner conversation.

Jane smiled as she made her way to a sturdy barstool. She looked to be in her twenties, not seventies, with perfect posture and her ponytail shining in the lamplight. "Why don't we go around the room first and introduce ourselves and say what we do," she began, engaging us individually and collectively in her mission.

After the introductions, Jane wove chimp stories and tales of her travels over the past year into a magnificent overview of her life and that of the chimps. She included parts of her childhood, when she was so engrossed with watching a hen lay an egg that her mother thought she was lost, only to discover her later covered with hay. Old Flo and David Greybeard resurfaced as she spoke of early days at Gombe, and then she led on to her recent talks in China and Spain and how her youth education program, Roots & Shoots, is succeeding around the world. Afterward, we all felt reinvigorated about modeling good stewardship for Mother Earth. An author sitting next to me looked speechless, glancing over at me and saying simply, "Wow!"

Jane's life story has encouraged others to start new endeavors and realize their dreams. Though she worked hard and with great focus to fulfill her childhood wish to study animals, I'm sure she never imagined that some of the doors she opened would lead to worldwide fame and influence.

Jane's New Mission

In fact, Jane's original mission transformed and broadened in the early 1990s, when she flew over Gombe Stream National Park and saw the massive deforestation of the land surrounding it. Peering down, she could see the well-defined green Gombe forest surrounded by dry, eroded soil. She was alarmed at the changed landscape and at that moment made a difficult decision that would fundamentally change her life. As much as she hated leaving her family of chimps and people, she realized she must also work to protect the environment around Gombe. Only then would the chimpanzees have a chance of surviving the encroachment of desperate humans who also need to survive in the same region.

On our trip, Tommy and I witnessed how Jane has tackled both crises with her grassroots organization, TACARE, which helps villagers convert traditional farms to farms that use sustainable methods. A German-born agriculturist, George Strunden, was instrumental in the design and organizing of the program. We visited the sites surrounding the national park and saw some of the successes with healthier crops and the beginnings of land restoration.

Jane and Mary Lewis, her longtime assistant, suggested we visit the villages participating in the TACARE program. Had we not included this "field trip," we would have missed a crucial part of the picture of Gombe's future.

A dedicated TACARE employee, Fadhili, met Tommy and me in Kigoma. Fadhili was very articulate and gracious as he drove us to some of the villages participating in the TACARE conservation efforts funded by the Jane Goodall Institute to address environmental issues as well as the prosperity of the Tanzanian people. Jane and Mary knew we'd be interested in seeing the environmental work being done around Gombe Stream National Park.

Jane was promoting a grassroots effort led by villagers to restore local vegetation and prevent erosion, encouraging a return of the verdant landscape that was thriving when she first arrived in Tanzania in 1960. Although Gombe Stream National Park had been protected from destruction, the land around it had not been. Because of wars and political upheavals, the hills and valleys along Lake Tanganyika from Kigoma and far north to Burundi supported refugees from the Congo and Rwanda until the early 1990s. Local Tanzanians also found the habitat good for establishing small farms, which required the cutting of trees. The resulting erosion had taken its toll. The corridors for migrating chimps and other animals—who require these undisturbed natural passages to travel and mate with animals in other regions, widening the gene pool—were being destroyed. The future looked very grim for chimpanzees and other species.

Several years before this, Jane studied the basic problem and designed an approach that would involve input from the local people. Each village near the park and farther north would choose a leader for the environment, a leader for education, and a leader for public health. Small villages might have one leader assume all three roles. They would communicate and work with one another and with workers from TACARE in Kigoma, and together begin to restore and maintain the land along the lake. As an incentive, a bicycle or piped-in water from fresh springs was supplied to each participating village. Tree farms were established so that firewood could be taken from those specific areas rather than by stripping the land everywhere. As a result, Lake Tanganyika is clear and erosion is receding for miles up and down the shores abutting Gombe National Park.

Fadhili parked the car and walked with Tommy and me to a clearing, where we could see for miles. The tiny green dots on the distant hillside represented new trees that had been planted a few years earlier. I raised my eyebrows and smiled as I told Fadhili, "I noticed that Lake Tanganyika is clear and you can see lots of fish swimming. I was told by health officials that there is no bilharzia."

"Yes!" Fadhili exclaimed, "This is the work of TACARE. And the people out here in the villages benefit as much as the fishermen and people living along the lake. You can look out across those hills," he said, pointing across the valley, "and see the small trees beginning to grow back. Clean water flows to many villages, and people seem empowered and happy with the results."

Jane also got involved in the coffee industry by supporting the local transition from traditional coffee growing to shade-grown coffee, which is more sustainable and better for the environment. Local Tanzanians run the coffee production in what is truly a success story—a win-win situation that provides work for local people while it benefits the land.

Tommy and I were invited to sit in on a meeting involving the managers—all Tanzanians—of the various coffee farms. The managers looked serious enough to be the US Senate. Asked to introduce us to the group, I started in Swahili, which they seemed to appreciate, and briefly described our lives. I relied on Fadhili to translate the rest. The most interesting moment came when I mentioned that, because of my own interest in land preservation, I had campaigned for Barack Obama. As soon as they heard this in translation, they burst into applause and cheers. It was a real icebreaker.

We then visited a coffee-growing farm and several of the small villages in TACARE's program. We also toured the new tree farms planted nine years earlier, which were thriving. Finally, we learned more about Jane's Roots & Shoots organizations at the local schools and villages. Roots & Shoots clubs have been successful at educating young people worldwide about the environment and inspiring children to participate in environmental preservation. The efforts of TACARE and Roots & Shoots are changing the lives of the people in this area while also sustaining and improving the local environment.

When we returned to our hotel that evening, I felt like a diplomat who had just toured a new project, and I thanked Fadhili for his time and his commitment to the organization. I was awed by Jane's resolve in tackling such a complex environmental challenge and working from the ground up to address it. After our return home, I sent a donation to TACARE in Fadhili's name to help support their efforts.

Recently, Wendy and I heard Jane speak to a large crowd in Southern California. Even though I was familiar with one of her key messages, it helped to hear it again. Jane quoted her mother, who told her when she was a young girl, "If you work hard, take advantage of opportunities, and never give up hope, you can accomplish your goals." At the end of the talk, Jane answered a high school student's question about career choice and suggested, "When the time is right, take the opportunity and go with it." In her own life, Jane certainly did.

To keep up with Jane would be impossible. Her pace and focus have always been like no one else's. As I reflect again on her genetics—with her race-car-driving father and her patient and focused mother—I see how some of her success might have started, but I believe it also took a strong, consistent upbringing to maximize those genes—nature and nurture beautifully merged.

So as I plunged back into my medical practice after returning from Gombe with Tommy, I thought about how lucky I was to have been one of Jane's students. Because of Jane I understood that we all remain students throughout our lives. I thought about her drive and her confidence in communicating with people about our current race to save species and the natural world around us, and I felt grateful that over the years she instilled in me a strong commitment to help in preserving our planet. This sense of purpose has kept me writing this book, even when it seemed like an impossible task.

As I continue to refine my own purpose in my community and family, I can still picture the image I formulated in a remote Tanzanian forest, a vision of our early ancestors making their way through the trees and undergrowth, foraging for food and raising their offspring. I picture how successful they were and the hardships they endured. When life seems difficult or the future of our beautiful earth appears to be compromised, my hope comes from my connection to this image. It also comes from knowing Jane.

EPILOGUE

In my sixty-first year, I made a brief escape from my busy life at work and at home to write the last chapter of this book. The week at my office had been especially stressful, with more patients than usual requiring urgent attention, but now I had twenty-four precious hours for my retreat to a cabin on Whidbey Island, two hours from Seattle.

Before settling down to write, I dropped off my laptop at the cabin and drove a few miles to Ebey's Landing, where a trail leads up to a bluff overlooking the Salish Sea. For the past thirty years I'd paid tribute to this unique nature reserve by hiking along the bluff trail, which separates rich farmland with views of the Olympic Mountains from a 150-foot drop-off to the water. Passing ships and a distant horizon would keep my eyes focused west, toward the sea, for most of the hike.

I wanted to hear the echoing call of the great blue heron and the splashing of waves on the sandy beach below. I wanted to remember the Native Americans who lived on this land not long ago. On this unusually warm April evening, I felt invigorated to be on the trail breathing the salty mists. I scampered down the steep hillside to focus on the waves and the setting sun and appreciate the efforts that have kept this land protected. The Ebey's Landing National Historical Reserve is a unique mix of federal, state, county, and private lands, all protected under a land trust. The concerted efforts of local and national groups prevailed in keeping this precious coastal habitat free from home construction so that trail users can continue to enjoy the land and its history.

As I walked, my attention drifted to the wispy clouds forming near the horizon and the deepening colors of the sun setting over the Strait of Juan de Fuca, reminiscent of the blazing sunsets over Lake Tanganyika. As they so often did, my thoughts turned to the Gombe chimps, especially Fifi's brother Figan, whose memory always tugged at my heart. In the wild, this proud male was three times stronger than any human his size, yet what good was Figan's power in the face of human encroachment on his land? Figan was, like all chimps, helpless against human weaponry—no match for gun-wielding poachers roaming the forests. Figan did not leave a single footprint of his life at Gombe. He was the ultimate environmentalist. It seemed unfair that his species might have limited time left, in large part due to the actions of humans.

A chimp's brain cannot complicate things quite like a human brain can. Although only 4 percent of our DNA differs from chimp DNA, it is that 4 percent that makes us human. Complex problem-solving and language capabilities of the human brain arise from that same small difference in DNA. It's that same 4 percent that pushes me, for example, to better understand our interconnectedness with chimps and our capacity as humans to respond more effectively to the challenges of our complicated world. If the chimps have been able to adapt and survive in their wild communities for so long, then surely we humans with our more complex brains should be able to forge a future that is sustainable for all of Earth's species.

Walking this trail so far from Africa and my youth, I remembered the first chimpanzee to steal my heart—Babu. Like Figan, Babu showed off his acrobatic skills for me, but unlike chimps in the wild, Babu often clung tightly to my chest for reassurance when he was frightened. I have always been haunted by the image of Babu before his rescue—an orphaned infant in a small crate in a West African market waiting to be sold as meat. In 1974, however, when I returned to California, I'd happily observed four-year-old Babu interacting with other chimps at the Stanford Outdoor Primate Facility. An older female chimp, Bashful, had "adopted" Babu and provided love and affection as though he were her own offspring. Watching Babu nurtured by his own kind was deeply reassuring.

Circumstances had changed for Babu four years after entering the primate facility. Babu and other chimps at the facility contracted hepatitis B from a new chimp member and became carriers. Babu, a younger female named Topsy, and Mowgli, the chimp who had infected them, were soon transferred to the National Institutes of Health in Maryland to be used in research to develop a cure for the disease they all carried.

Babu was then transferred to many different facilities, first in Texas, then New Mexico, then back to Texas, where he and Topsy had two offspring together, and finally Colorado. Though I was troubled to hear of Babu's being used as a medical research subject, I took solace in the fact that his rescuers—the California couple from Woodside—contacted each facility Babu entered to assure his well-being. They even visited Babu in Texas. It was certainly not their choice for him to become a lab primate, but it was out of their control. After Babu's transfer to Colorado, my marvelous first chimpanzee companion died from complications related to the research. Babu was twenty-nine years old.

Babu's journey spanned continents, embraced human and chimp caretakers, and contributed to hepatitis B research in a study that ultimately caused his death. When I received news of his final years, I shed tears just as I did during my last visit with him in 1973 when he was two and a half years old. I can still picture him leaping into my arms and hugging me longer than usual on our final get-together before I left for Africa.

At our human hands, Babu had experienced both the joy of rescue and connection and the suffering of capture and medical experimentation. Our bond with these primate cousins is complicated and finally evolving. For many years, we've used chimpanzees solely for our own purposes—whether as human replacements in the space race or as lab subjects in our medical research, but in the past decade, we've discovered that with modern technology, there are many ways to do medical research without using chimpanzees.

In the United States, there's some good news: in June 2013, the National Institutes of Health announced that it would release approximately 360 chimps that had been locked in cages and used for biomedical experiments for up to thirty years. They were no longer considered crucial for research. Most had been kidnapped from the wild as infants, just like Babu, and brought here to spend the rest of their lives in a laboratory. Fifty were still held for medical purposes, making the United States the only developed country still using chimps for research—until June 2015, when even these chimps began their preparation for release into sanctuaries.

Now a chimp that has lived his life in a small metal cage as the subject of scientific studies can finally be released to an outdoor sanctuary like Chimp Haven in Louisiana. There are many poignant videos on YouTube that show these chimps and their first taste of freedom (search "laboratory chimps released"). Hesitantly, they venture out of their cages into these sanctuaries surrounded by trees, sky, and other chimps for the very first time. It's astounding to see their curious—but also fearful—facial expressions while cautiously entering the outdoors and being able to embrace one another for comfort. My heart holds a special place for many of the chimps that still remained caged for research as of the writing of this book as they await a spot in suitable sanctuaries. There was no need for them to continue to suffer in captivity for medical experiments. I hope one day to hear that they too have been given freedom, surrounded by their own kind and those tireless and compassionate human caretakers who work in chimpanzee sanctuaries.

My walk along the Whidbey Island bluff trail also took me back to my trekking days with Hamisi Matama as we followed the chimps together. I heard from Tony that Hamisi was once again working as a field assistant/researcher at Gombe and was known as the "wise man of the forest" because of his knowledge of the species names and medicinal properties of local plants. According to Tony, Hamisi also knew "many tales of former times now all but forgotten by young people busy with schooling and more interested in cell phones, laptops, and football." Of interest to me was another fact I learned from Tony: Hamisi's traditional name was Mlongwe. He didn't use this name while I was at Gombe, but people at the camp now knew him by it.

Hamisi was continuing his work at Gombe as a chimp researcher and medicinal plant educator as I entered my thirty-second year in the same clinic in Seattle, still excited to connect with those who sought my help. I still enjoy the challenge of diagnosing and designing treatment plans. The passion that keeps me going through the twelve-hour days is the adventure of stepping into the interesting lives of my patients and developing trust and mutual respect in our journey together. To be a part of a patient's life has always seemed an extraordinary privilege. It's very similar to my feelings about the Gombe chimps—I feel gratitude for having been allowed to observe them and be immersed in their lives. My experience with the chimps enlarged my understanding of the world and of our species. I believe it made me a better doctor and a better father. I know it made me a better human being.

I am content to be alive now with modern conveniences, such as hot water and a bathtub to relax me at the end of a hard day, but I often find myself wanting more of a connection to nature, more time to build stronger connections with friends and family, and to feel just physically tired instead of emotionally exhausted at the end of the day.

As I look ahead, I often think about a sojourn in Africa, Asia, or Central America, perhaps for months or even years. My wife could teach and I could continue to practice medicine. Our boys—now young men—could visit us from time to time and explore another new culture. Wherever I live, I want to feel a strong sense of purpose. Because of my time with the chimps, I hope that will include helping preserve ecologically rich places such as Gombe or the rainforests of Brazil or Madagascar. I also want to be able to hear stories of other people's lives and share some of my own past adventures.

I will be forever grateful to Jane, to the chimps of Gombe, and to Hamisi for giving me the experience of living in a wild forest, an experience that helped me build my hopes and dreams and gain self-confidence. Wendy and I dream that human primates will learn to live in harmony with nature as the chimps do. And we dream that our grandchildren and great-grandchildren will enjoy the profound pleasures of encountering a primate cousin—a distant family member who still has so much to teach us about being human.

ACKNOWLEDGEMENTS

A huge thanks to Jane Goodall for mentoring me throughout my adult life and writing the foreword to this book.

My parents Patty and Jack carefully saved the volumes of letters I wrote to them from Africa, which became the bases for part 1 of this book.

Alyssa O'Brien, Stanford lecturer and writer (*The Quilt of My Life*) inspired me in the early stages of writing *Following Fifi* at Stanford's first "The Write Retreat" for alumni.

Michelle Tessler was my trusted and successful literary agent who found Claiborne Hancock, head of Pegasus, to take on *Following Fifi* with enthusiasm and grace.

Laura Garwood was exceptional in her polishing and editing of the manuscript.

ACKNOWLEDGMENTS

Six authors provided me with guidance in shaping and refining the manuscript:

Thor Hanson (*The Triumph of Seeds*)

Brenda Peterson (*Wolf Nation*)

Clare Hodgson Meeker (*Rhino Rescue!*)

Peter Ames Carlin (*Bruce*)

James Thayer (*House of Eight Orchids*)

Phil Hanrahan (*Life After Favre*)

Four free-lance editors supported me: Linda Gunnarson, Marlene Blessing, Michele Rubin, and Cypress House.

Contributors to the manuscript:

Hamisi Matama Majana, field assistant and friend

David Anthony Collins (Anton), baboon researcher and translator

Abdul Ntandu, field assistant and translator

Thomas Crocker, son

Former Stanford and Gombe researchers:

Nancy Merrick, internist

Chuck de Sieyes, family physician

Kathryn Morris, family physician

Other contributors:

Richard Wrangham, Harvard Department of Human Evolutionary Biology (*Demonic Males*),

David A Hamburg, former president, National Institute of Medicine, Mary Lewis,

Paul Witt, Nanette Leuschel, Emily Polis Gibson, Grant Scull, Travis Abbott, Wayne Dodge, Tobias Dang, Susan Crocker, and Hank Klein.

The Jane Goodall Institute and TACARE.

ACKNOWLEDGMENTS

Photo support: Mary Paris of the Jane Goodall Institue, National Geographic, Curt Busse, Emilie Riss, Anne Pusey, Grant Heidrich, Maria Fernandez, and Aadje Geertsema for her cover photo.

Finally to my wife Wendy for endless hours of supporting my writing, my son Patrick for his interest in the chimps and perfecting their calls, and my son Thomas (Tommy) for venturing back to Gombe with me.

It does take a village and I appreciate all the support I received in writing my first book, *Following Fifi*.

INDEX